ONE STOP DOC

Musculoskeletal System

One Stop Doc

Titles in the series include:

Cardiovascular System – Jonathan Aron
Editorial Advisor – Jeremy Ward

Cell and Molecular Biology – Desikan Rangarajan and David Shaw
Editorial Advisor – Barbara Moreland

Endocrine and Reproductive Systems – Caroline Jewels and Alexandra Tillett
Editorial Advisor – Stuart Milligan

Gastrointestinal System – Miruna Canagaratnam
Editorial Advisor – Richard Naftalin

Nervous System – Elliott Smock
Editorial Advisor – Clive Coen

Metabolism and Nutrition – Miruna Canagaratnam and David Shaw
Editorial Advisors – Barbara Moreland and Richard Naftalin

Renal and Urinary System and Electrolyte Balance – Panos Stamoulos and Spyridon Bakalis
Editorial Advisors – Alistair Hunter and Richard Naftalin

Respiratory System – Jo Dartnell and Michelle Ramsay
Editorial Advisor – John Rees

ONE STOP DOC

Musculoskeletal System

Wayne Lam BSc(Hons)
Fifth year medical student, Guy's, King's and
St Thomas' Medical School, London, UK

Bassel Zebian MBBS BSc(Hons)
GKT Graduate and Pre-Registration House Officer in General Medicine,
Medway Maritime Hospital, UK

Rishi Aggarwal MBBS
Senior House Officer in General Medicine,
Queen Elizabeth Hospital, London, UK

Editorial Advisor: Alistair Hunter BSc(Hons) PhD
Senior Lecturer in Anatomy, Guy's, King's and
St Thomas' School of Biomedical Sciences, London, UK

Series Editor: Elliott Smock BSc(Hons)
Fifth year medical student, Guy's, King's and
St Thomas' Medical School, London, UK

Hodder Arnold

A MEMBER OF THE HODDER HEADLINE GROUP

First published in Great Britain in 2005 by
Hodder Education, a member of the Hodder Headline Group,
338 Euston Road, London NW1 3BH

http://www.hoddereducation.co.uk

Distributed in the United States of America by
Oxford University Press Inc.,
198 Madison Avenue, New York, NY10016
Oxford is a registered trademark of Oxford University Press

Whilst the advice and information in this book are believed to be true and
accurate at the date of going to press, neither the authors nor the publisher
can accept any legal responsibility or liability for any errors or omissions
that may be made. In particular, (but without limiting the generality of the
preceding disclaimer) every effort has been made to check drug dosages;
however it is still possible that errors have been missed. Furthermore,
dosage schedules are constantly being revised and new side-effects
recognized. For these reasons the reader is strongly urged to consult the
drug companies' printed instructions before administering any of the drugs
recommended in this book.

British Library Cataloguing in Publication Data
A catalogue record for this book is available from the British Library

Library of Congress Cataloging-in-Publication Data
A catalog record for this book is available from the Library of Congress

ISBN-10: 0 340 88505X
ISBN-13: 978 0 340 88505 5

1 2 3 4 5 6 7 8 9 10

Commissioning Editor: Georgina Bentliff
Project Editor: Heather Smith
Production Controller: Jane Lawrence
Cover Design: Amina Dudhia
Illustrations: Cactus Design
Index: Indexing Specialists (UK) Ltd

Hodder Headline's policy is to use papers that are natural, renewable and recyclable
products and made from wood grown in sustainable forests. The logging and manufacturing processes
are expected to conform to the environmental regulations of the country of origin.

Typeset in 10/12pt Adobe Garamond/Akzidenz GroteskBE by Servis Filmsetting Ltd, Manchester
Printed and bound in Spain

What do you think about this book? Or any other Hodder Education title?
Please visit our website at **www.hoddereducation.co.uk**

CONTENTS

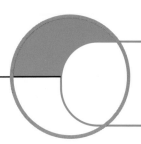

PREFACE

From the Series Editor, Elliott Smock

Are you ready to face your looming exams? If you have done loads of work, then congratulations; we hope this opportunity to practice SAQs, EMQs, MCQs and Problem-based Questions on every part of the core curriculum will help you consolidate what you've learnt and improve your exam technique. If you don't feel ready, don't panic – the One Stop Doc series has all the answers you need to catch up and pass.

There are only a limited number of questions an examiner can throw at a beleaguered student and this text can turn that to your advantage. By getting straight into the heart of the core questions that come up year after year and by giving you the model answers you need this book will arm you with the knowledge to succeed in your exams. Broken down into logical sections, you can learn all the important facts you need to pass without having to wade through tons of different textbooks when you simply don't have the time. All questions presented here are 'core'; those of the highest importance have been highlighted to allow even sharper focus if time for revision is running out. In addition, to allow you to organize your revision efficiently, questions have been grouped by topic, with answers supported by detailed integrated explanations.

On behalf of all the One Stop Doc authors I wish you the very best of luck in your exams and hope these books serve you well!

From the Authors, Wayne Lam, Bassel Zebian and Rishi Aggarwal

The aim of this book is to review and simplify information concerning the musculoskeletal system in a question and answer format. This book covers the principles of musculoskeletal physiology and anatomy, as well as some biochemistry and pharmacology that are relevant to your future clinical studies. It gives you an opportunity to have a quick tour of all the important topics concerning the musculoskeletal system and gives you exam experience.

In this book, we have also tried to highlight some key questions which concern the basic principles of the topic. Some related clinical scenarios have also been discussed. We found the musculoskeletal system to be a very challenging aspect of medicine and we hope that this book will provide a complete and simplified review for your learning.

From the Author, Wayne Lam
Many thanks to my parents for being the best parents in the world. I would also like to thank my brother (Tim) for all his practical jokes to cheer me up during the long writing sessions, and Ami who made me tea and coffee to keep me awake.

From the Author, Bassel Zebian
To my father, mother, brother and sister – I am eternally grateful for your continuous support over the years. Many thanks to Wayne and Rishi for all the hard work you put in. Thank you Tash for being there when it counted. Finally, thank you Miss Barnes for all your help.

From the Author, Rishi Aggarwal

I would like to thank my good friend Bassel who asked me to become an author in the first place. I would also like to mention my parents who will no doubt boost sales by letting everyone know that their son is now an author. Finally, thank you to my brother (Rupesh) and sister (Roshni) for providing laughter during the long sessions of writing.

Most of all, we would all like to thank the real brain box behind the book, Dr Hunter. He kindly supervised every stage of the project with great patience. Without him this book would not have been possible. We would like to thank Elliott for letting us participate in a project that is sure to be very successful. Thanks everyone at Hodder Arnold Health Sciences Publishing (especially Heather) for putting in a great amount of time and effort in bringing the book together.

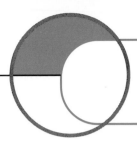

ABBREVIATIONS

ACh	acetylcholine
ATP	adenosine triphosphate
Ca^{2+}	calcium ion
CN	cranial nerve
GP	general practitioner
K^+	potassium ion
Na^+	sodium ion
PO_4^{3-}	phosphate

OVERVIEW OF THE MUSCULOSKELETAL SYSTEM

1. Complete the following diagrams with the options provided

Options

A. Posterior   **B.** Proximal   **C.** Eversion
D. Medial   **E.** Rotation   **F.** Lateral
G. Superior   **H.** Abduction   **I.** Distal
J. Anterior   **K.** Opposition   **L.** Adduction
M. Inversion   **N.** Inferior   **O.** Coronal plane
P. Horizontal plane   **Q.** Median plane   **R.** Sagittal plane

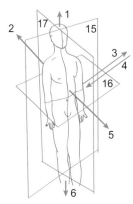
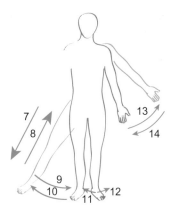

2. Concerning the connective tissues

     a. 20 per cent of the body is made of connective tissues
     b. Hyaline cartilage is found in intervertebral discs
     c. Fibroblasts in fibrocartilage give it its flexible characteristics
     d. Fibrocartilage is the major connective tissue in the pinna of the ear
     e. Chondrocytes are the cells of cartilage

3. What is the extracellular matrix? What is its function?

4. Describe the differences between tendons, fascia and ligaments

EXPLANATION: OVERVIEW OF THE MUSCULOSKELETAL SYSTEM

All descriptions in human anatomy are expressed in relation to the **anatomical position**:

- **Anatomical planes:** the **median plane** is a vertical plane passing through the body from front to back longitudinally. **Sagittal planes** are vertical planes parallel to the median plane. **Coronal planes** are vertical planes at right angles to the median plane, while **horizontal planes** pass through the body at right angles to both the median and coronal planes.
- **Terms of relation: anterior** is nearer to the front, **posterior** is nearer to the back. **Superior** is nearer to the head, **inferior** is nearer to the feet. **Medial** is nearer to the median plane, and **lateral** is farther from it
- **Terms of comparison: proximal** is nearest to trunk, while **distal** is farther from it. **Superficial** means nearer to the surface, while **deep** is farther from it. **External** means toward or on the exterior, and **internal** means towards or in the interior. **Ipsilateral** means on the same side, while **contralateral** means on the opposite side of the body
- **Terms of movement: flexion** indicates bending, and **extension** is straightening of body parts. **Abduction** is the movement away from the median plane, whereas **adduction** moves toward the median plane **Opposition** is the movement of the thumb to another digit. **Rotation** is the turning of a body part around its long axis. **Eversion** of the foot means moving the sole away from the median plane. **Inversion** indicates the movement of the sole toward the median plane.

Connective tissues are supporting tissues containing **extracellular matrix** and cells. The extracellular matrix is made of collagen, elastins and 'ground substance'. They make up about **70** per cent of body mass. They function to hold organs together and may degenerate with age, hence they are involved in many disease processes (3). Generalized connective tissues include **fibroblasts** (present in fascias, tendons and ligaments). **Cartilage** is a special connective tissue, containing **chondrocytes** which control the extracellular matrix. It is divided into three types:

- **Hyaline cartilage:** is found in most synovial joint surfaces and anterior ends of the first to tenth ribs
- **Fibrocartilage:** can be found in intervertebral discs. It contains **collagen**, making it flexible with a high tensile strength
- **Elastic cartilage:** contains **elastic fibres**. It can be found in the pinna of the ear, nose and larynx.

Tendons consist of thick collagen fibres parallel to the direction of pull, connecting muscles to bones. **Fascia** is tendon-like connective tissue arranged in sheets or layers. **Ligaments** are collagen fibres connecting bones to one another (4).

Answers
1. 1 – G, 2 – A, 3 – F, 4 – D, 5 – J, 6 – N, 7 – I, 8 – B, 9 – L, 10 – H, 11 – M, 12 – C, 13 – H, 14 – L, 15 – O, 16 – P, 17 – Q or R
2. F F F F T
3. See explanation
4. See explanation

5. Name four major functions of bone

6. Concerning bones in the human body

    a. Intramembranous ossification is the development of bone from the condensation of mesenchyme in the prenatal period

    b. In endochondrial ossification, cartilaginous tissue derived from mesenchyme is replaced with bone within sites called ossification centres

    c. Trabecular compact bone is a network of bony threads arranged along the lines of stress within the bone cavity

    d. Haemopoesis takes place within the bone cavity

    e. Osteoclasts erode bone

7. The following diagram shows a long bone. Label it with the options provided

Options

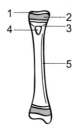

    A. Metaphysis   **B.** Articular cartilage
    C. Diaphysis   **D.** Apophysis
    E. Physis   **F.** Epiphysis

8. Concerning the development of a long bone, put the following statements in chronological order

Options

    A. Growth of blood vessels accelerates through the periosteum and bone collar, forming the primary ossification centre at the centre of the diaphysis

    B. Development of osteoprogenitor cells and osteoblasts. The perichondrium becomes a periosteum in the mid-shaft of the diaphysis

    C. Establishment of secondary ossification centres in the centre of each epiphysis

    D. The developing cartilage model assumes the shape of the bone to be formed

    E. A network of bony trabeculae spreads out and links up with previously formed bone collar

    F. Formation of cortical bone of the diaphysis, with the epiphysis still composed of cartilage

    G. The development of chondroblasts in primitive mesenchyme, forming the perichondrium and cartilage

---

Ca^{2+}, calcium ion; PO_4^{3-}, phosphate

EXPLANATION: BONES

Bone functions to (i) provide shape, support and levers for movements, (ii) protect internal organs, (iii) store the body's Ca^{2+}, and (iv) produce blood cells (haemopoiesis) (5). Bones may be developed in two ways:

1. In **intramembranous ossification**, bones develop from the condensation of mesenchyme in the prenatal period

2. **Endochondral ossification** is the ossification of the pre-existing hyaline cartilage. The process starts at the **primary ossification centre**, which is located at the diaphysis of the long bone (area between two ends of the bone). Here, the cartilage cells increase in size. The matrix formed becomes calcified and the cells die. At the same time, deposition of a layer of bone under the **perichondrium** (which surrounds the diaphysis) and becomes the **periosteum**. Vascular connective tissues derived from the periosteum breaks up the cartilage, creating spaces that fill with haemopoietic cells. This process continues towards the epiphyses (ends of the bone). The **epiphyseal growth plate** (diaphyseal-epiphyseal junction) is the predominant site of longitudinal growth of the bone. At birth, **secondary ossification centres** appear in the epiphyses, where osteoblasts continue to ossify cartilage so the bone grows longer.

Bone is a special connective tissue, composed of microscopic crystals of calcium phosphate within a collagen matrix. It is highly vascular, and is surrounded by the periosteum. Bones are hollow. The cavity is filled with bone marrow which produces blood cells. Lamellar bone within the marrow cavity presents as a network of bony threads, arranged along the lines of stress termed the **trabecular compact bone**. Bone surrounding the cavity is organized into compact layers, and this region is called the **compact lamellar bone**.

There are two main patterns of bone. **Woven bones** have a haphazard organisation of collagen fibres and are mechanically weak. **Laminar bones** have a regular parallel alignment of collagen in sheets and are mechanically strong (8).

Osteocytes are inactive cells of the bone. They are surrounded by mineralized **osteoid**, giving it the property of rigidity and strength while retaining its elasticity. Bones remodel throughout life in response to mechanical demands. **Osteoblasts** are bone-forming cells, and **osteoclasts** erode bone by the process of reabsorption.

Answers
5. See explanation
6. T T T T T
7. 1 – B, 2 – F, 3 – A, 4 – D, 5 – C
8. 1 – G, 2 – D, 3 – B, 4 – A, 5 – E, 6 – F, 7 – C

9. Concerning joints

a. Sutures are fibrous joints
b. The interosseous membrane between the radius and ulna is an extended fibrous tissue of a fibrous joint
c. Fibrocartilage covers the bone in a primary cartilaginous joint
d. The pubic symphysis is a cartilaginous joint
e. Intervertebral joints are examples of synovial joints

10. Concerning the synovial joints

a. Cartilage of a synovial joint is supplied by a rich neurovascular network
b. Joint capsules are involved in proprioception
c. Synovial fluid is secreted by the synovial membrane to reduce resistance upon movements of the joints
d. Hinge joints allow biaxial movements
e. Ball and socket joints allow multiaxial movements

11. The following diagram shows a synovial joint. Label the diagram with the options provided

Options

A. Joint cavity   **B.** Articular cartilages
C. Synovial membrane   **D.** Periosteum
E. Fibrous capsule

12. The following diagram shows different types of synovial joints. Label them with the options provided

Options

A. Saddle joint   **B.** Ball and socket joint
C. Hinge joint   **D.** Condyloid joint
E. Pivot joint   **F.** Plane joint

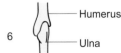

EXPLANATION: JOINTS

Joints are articulations between bones, a bone and a cartilage, or between cartilages. Three types of joints include:

- **Fibrous joints**: articulations are united by **fibrous tissues**. An example is a joint between the flat bones of the cranial vault, where they are known as **sutures**. **Gomphoses** are fibrous joints between the teeth and the jaw, while the **interosseous membrane** between the radius and ulna is an extended fibrous tissue of a fibrous joint

- **Cartilaginous joints**: in **primary cartilaginous joints**, bones are joined together by **hyaline cartilage**, usually a temporary union of bones. The **epiphyseal cartilaginous plate** separating the epiphyses and diaphysis is an example. **Secondary cartilaginous joints** or **symphyses** occur only in the median plane. The articulating surfaces are covered with hyaline cartilage and unite bones by strong fibrous tissues. Examples include the **intervertebral joints** and the **pubic symphysis**

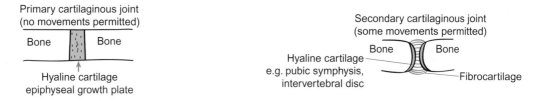

- **Synovial joints**: these highly mobile joints have three special features (see figure for question 11):
  - Each of the bones involved is usually coated with a layer of **hyaline cartilage**. The cartilage has no nervous or blood supply, and relies on its nourishment by the surrounding **synovial fluid**
  - **Joint capsules** are present in synovial joints. They are lined with **synovial membrane**, which secretes lubricating synovial fluid. These capsules contain sensory nerve endings, providing the brain with information concerning movement and position of the joint and the body (**proprioception**)
  - The synovial joint contains a **joint cavity**. These joints are usually stabilized by associated ligaments and muscles.

There are six types of synovial joints. They are (i) pivot joints, (ii) ball and socket joints, (iii) condyloid joints, (iv) plane joints, (v) saddle joints and (vi) hinge joints (see figures for question 12).

Answers
9. T T F T F
10. F T T F T
11. 1 – C, 2 – B, 3 – A, 4 – E, 5 – D
12. 1 – E, 2 – B, 3 – D, 4 – F, 5 – A, 6 – C

13. The following diagram shows the structure of part of a skeletal muscle. Label it with the options provided

Options

A. Perimysium   **B.** Myofibrils
C. Sarcolemma   **D.** Epimysium
E. Fasciculi   **F.** Endomysium
G. Muscle fibres

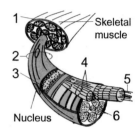

14. The following diagram shows the appearance of a human skeletal muscle under the electron microscope. Label it with the options provided

Options

A. J-band   **B.** A-band
C. H-band   **D.** Z-line
E. M-line   **F.** Actin
G. Myosin

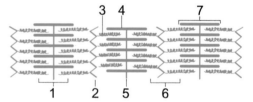

15. What happens to the H-band, I-band and A-band of a sarcomere during muscle contraction? Choose the best answer from the options below

A. The width of the I-band is decreased   **B.** The width of the A-band is decreased
C. The width of the H-band is increased   **D.** All of the above occur
E. None of the above occur

16. Define the terms isometric contraction and isotonic contraction

17. The following 'length–tension' curve of a single muscle has been obtained

A. What does the active curve indicate?
B. What does the passive curve indicate?
C. Which point in the diagram shows no overlap between the majority of the muscle's thick and thin filaments?

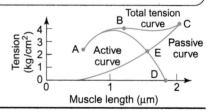

EXPLANATION: SKELETAL MUSCLES AND MUSCLE CONTRACTION AT CELLULAR LEVEL

Skeletal muscles are muscles attached to bones and cartilage and are controlled by the somatic nervous system. Skeletal muscle is the functional contractile unit, responsible for voluntary movement. Skeletal muscles are composed of a collagenous connective tissue framework and muscle fibres supplied by a neurovascular bundle.

Muscle fibres (or cells) are grouped together into **fasciculi**, with the **endomysium** occupying spaces between individual muscle fibres as supporting tissues. Each of the fasciculi is surrounded by the **perimysium**, a loose connective tissue. Fascicles are grouped together to form a muscle mass by the **epimysium**, a dense connective tissue (see figure in question 13).

Within each muscle fibre there are two types of proteinous filaments:

- Thin filaments: **actin**, **tropomyosin**, and **troponin**
- Thick filaments: **myosin**.

Each functional contractile unit contains the above filaments, and is called a **sarcomere**. Sarcomeres present with a characteristic cross-striation pattern.

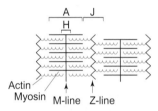

Muscle contraction depends upon the regularly repeating sets of sarcomeres, where actin interdigitates with myosin. The contractile mechanism in skeletal muscle depends on **cross-bridge** (bonding) interactions between these two filaments and the sequence is shown on the figure on page 18. During muscle contraction both the **I-bands** and the **H-bands** are shortened. There is no change in length of the **A-bands**.

An **isometric contraction** means that muscle force changes as it contracts at a constant length. An **isotonic contraction** means that muscle changes in length as it contracts against a constant load **(16)**. The sarcomere length (muscle length) can be related to the amount of tension a muscle can produce under isometric conditions by the **'length–tension' curve**.

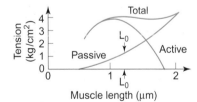

The **active curve** is a function of the number of cross-bridges available for cross-bridging **(17A)**. The **passive curve** is a function of the length of the relaxed muscle **(17B)**. The **total tension curve** is the sum of the active and passive curves.

At L_0 on the above diagram, alignment of actin and myosin is perfect, giving the maximum possible number of cross-bridges formed between actin and myosin, so every actin and myosin can cycle. **(17C)**

Answers

13. 1 – D, 2 – A, 3 – C, 4 – G, 5 – B, 6 – F, 7 – E
14. 1 – C, 2 – D, 3 – F, 4 – G, 5 – E, 6 – A, 7 – B
15. A
16. See explanation
17. See explanation; c – point E

18. Concerning the skeletal muscles

a. The sarcotubular system concerns the regulation of Ca^{2+} in muscle cells
b. The transverse tubules are continuous with the membrane of the muscle fibre
c. Ca^{2+} enters the myoplasm from the sarcoplasmic reticulum by active transport
d. A cotransporter system on the myoplasm maintains the low Ca^{2+} concentration at resting state
e. The plateau of an action potential helps to maintain the opening of the Ca^{2+} channels on the sarcoplasmic reticulum

19. Define twitch and tetanus

20. The following statements concern the sequence of contraction–relaxation in skeletal muscle. Put them in correct chronological order

Options

A. Action potential travels over the surface of the skeletal muscle cell and down the t-tubules
B. Cross-bridge cycling
C. Disconnection of actin–myosin cross-bridges
D. Ca^{2+} binds to troponin
E. Activation of dihydropyridine receptors in the t-tubular membrane
F. Muscle relaxation
G. Tropomyosin moves and exposes attachment sites for the myosin cross-bridges
H. Muscle contraction
I. Ca^{2+} stored in the sarcoplasmic reticulum is released into the intracellular compartment
J. Active transport of Ca^{2+} into the sarcoplasmic reticulum

21. The following graphs show the force–velocity relationship of a skeletal muscle

Options

A. Which graph suggests differences in the force–velocity relationship due to changes in myosin ATPase activity?
B. Which graph suggests changes in the force–velocity relationship of skeletal muscle due to changes in the number of motor units?

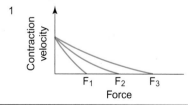

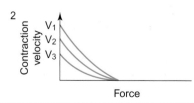

ATP, adenosine triphosphate; Ca^{2+}, calcium ion

EXPLANATION: SKELETAL MUSCLE CONTRACTION AT MOLECULAR LEVEL (i)

Ca^{2+} binds to **troponin** and uncovers the cross-bridge binding site on **myosin**. This allows **cross-bridges** on the myosin to attach to the thin filament during muscle contraction. Hence, the regulation of the cellular Ca^{2+} concentration is essential. This is controlled by the **sarcotubular system**.

The sarcotubular system consists of the **sarcoplasmic reticulum**, which surrounds the muscle fibres as a membrane. The system also consists of vesicles and transverse tubules (the **t-system**, which are tubules continuous with the membrane of the muscle fibre).

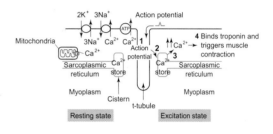

The sarcoplasmic reticulum stores a large amount of Ca^{2+}. Its membrane contains Ca^{2+}-**releasing** channels, which are closed when the surrounding cytoplasmic Ca^{2+} concentration is low but open when the surrounding Ca^{2+} concentration is high.

When an action potential (see page 15) is transmitted along the sarcolemma, the t-tubular membranes, covered by voltage sensors (**dihydropyridine receptors**), are briefly depolarized. This results in the opening of the Ca^{2+} channels in the sarcoplasmic reticulum membrane, and a pulse of Ca^{2+} is released from the sarcoplasmic reticulum into the **myoplasm** (the contractile portion of the muscle cell). Ca^{2+} then binds to **troponin** to initiate muscle contraction (see diagram above). However, Ca^{2+} also activates the ATP-driven sarcoplasmic reticulum pumps which restore the resting state unless stimulated by another action potential.

Due to the fact that Ca^{2+} ions are rapidly pumped back into the sarcoplasmic reticulum before the muscle gains sufficient time to develop its maximal force, such a response to a single action potential is termed a **twitch**. If twitches are repeated, this may lead to **tetanus**, where pulses are added together to maintain a saturated Ca^{2+} concentration for troponin in the myoplasm **(19)**. Here, all cross-bridges that can cycle with sites on the actin will be continuously cycling.

The **'force–velocity' relationship (21)** of skeletal muscle is shown below. It shows how much force can be produced if the muscle is allowed to shorten as it contracts, and is directly related to cross-bridge function. Note that:

- V_{max} = maximum speed of shortening, it:
  - Occurs when force is minimal
  - Reflects the maximum cycling rate of the cross-bridges
  - Is determined by the type of myosin that makes up the thin filament
- **Force** is minimal when muscle shortens rapidly and maximal when velocity = 0 (i.e. isotonic conduction).

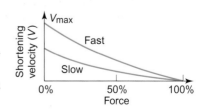

Answers

18. T T T T F
19. See explanation
20. 1 – A, 2 – E, 3 – I, 4 – D, 5 – G, 6 – B, 7 – H, 8 – J, 9 – C, 10 – F (see figure on page 18)
21. A – 2, B – 1

22. What is a motor unit?

23. What is a miniature endplate potential? Does it generate any action potentials?

24. Concerning the action potential illustrated below, which of the statements are true and which are false?

a. The interval between b and c is caused by a diffusion of Na⁺ into the cell

b. The interval between b and c is caused by an infusion of Na⁺ into the cell by an active transport system

c. The interval between c and d is due to the infusion of K⁺ into the cell by active transport

d. The interval between c and d is caused by the active transport of Na⁺

e. Another action potential may be triggered in the interval between f and g

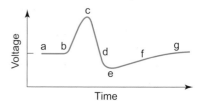

ACh, acetylcholine; Ca²⁺, calcium ion; Na⁺, sodium ion; K⁺, potassium ion

EXPLANATION: NERVOUS SIGNAL TRANSDUCTION

A **motor unit** is the combination of the motor nerve and the muscle fibres it innervates **(22)**.

To stimulate muscle contraction, a signal is passed from the motor nerve to the muscle by chemical transmission via the neuromuscular junction (neuromuscular transmission) see page 15. The event is triggered by an **action potential** (depolarization of the cell membrane). The action potential acts as a signal which propagates along the motor nerve.

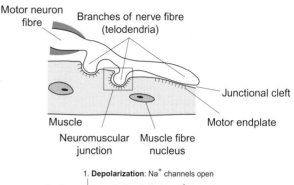

An action potential is produced by the **simple diffusion** of ions through channels on the cell membrane. **Depolarization** is via **Na$^+$** influx through voltage-gated channels. At the peak of the action potential, Na$^+$ conductance is at its maximum. At this point, the membrane potential is close to Na$^+$ equilibrium, therefore there is very little influx of Na$^+$ into the cell. However, at **repolarization**, K$^+$ **efflux** occurs through voltage-gated channels. Each depolarization is one stimulation, and a series of depolarizations is required if the muscle is to remain contracted. When an action potential reaches the neuromuscular junction, the events illustrated on the figure on page 15 occur to initiate muscle contraction.

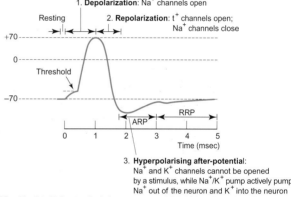

ARP = Absolute Refractory Period
RRP = Relative Refractory Period

Depolarisation needs to exceed a certain **threshold** to fire an action potential. If it is not reached, no action potential would occur (**all or nothing** law of action potential.) An increased intensity of a stimulus does not affect the intensity of an action potential.

Absolute refractory period of an action potential is the period during which a second action potential cannot be induced, regardless of how strong the stimulus is (due to voltage-inactivation of Na$^+$ channels). The length of this period determines the maximum frequency of action potentials. **Relative refractory period** is the period during which a greater than normal stimulus is required to induce a second action potential.

When the **ligand-gated Na$^+$ channels** open, the endplate region of the muscle membrane is depolarized. If a certain membrane potential threshold is reached, a muscle action potential is triggered from the endplate region, propagating away from the endplate, across the muscle surface and triggering muscle contraction. A **miniature endplate potential** results from the random release of a quantal package of ACh, producing a small depolarization of the postsynaptic membrane. It does not generate action potentials **(23)**.

Answers

22. See explanation
23. See explanation
24. T F F F T

25. Concerning the neuromuscular junction

a. Synthesis of ACh is catalysed by acetylcholinesterase
b. Far more ACh is released than is required to produce an endplate potential that is sufficient to trigger a muscle action potential
c. ACh binds postsynaptic nicotinic receptors leading to an influx of Na^+
d. Muscle action potential follows the all-or-nothing law
e. The number of nicotinic ACh receptors activated is proportional to the endplate potential

26. The following flow chart summarizes events occur during neuromuscular transmission. Please fill in the blanks with the options provided

Options

A. Release of ACh to the postsynaptic membrane where it binds to its receptors
B. Increased conductance to Na^+ ions
C. Depolarization of muscle membrane adjacent to endplate
D. Voltage-gated calcium channels open
E. Influx of extracellular Ca^{2+} ions into the axon terminal
F. Local depolarization of postsynaptic membrane (endplate potential)

1. Action potential travels down the axon to presynaptic motor axon terminal
2. _____
3. _____
4. _____
5. Opening of ligand-dependent channels
6. _____
7. _____
8. _____
9. Action potential spreads across the surface of skeletal muscle cell leading to muscle contraction

ACh, acetylcholine; Na^+, sodium ion; Ca^{2+}, calcium ion; K^+, potassium ion

EXPLANATION: NEUROMUSCULAR TRANSMISSION

The sequence of neuromuscular transmission at the neuromuscular junction is illustrated in the following diagram:

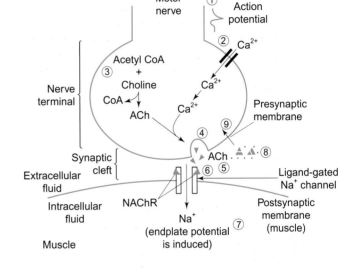

1. The action potential arrives at the presynaptic junction.
2. Conductance of Ca^{2+} ions is increased and there is increased influx of free extracellular Ca^{2+} ions.
3. Acetyl CoA + choline → production of ACh, a **neurotransmitter** (used to signal between the nerve and the skeletal muscle) (i) This process is catalysed by the enzyme choline acetyl transferase (CAT). ACh is stored in vesicles when not used and is protected from degredation.
4. The influx of Ca^{2+} ions acts as a signal for the release of ACh through the presynaptic membrane at the nerve terminal by the process of **exocytosis**. A larger amount of ACh is released than required to ensure the production of an end-plate potential sufficient enough to trigger a muscle action potential.
5. Acetylcholine diffuses through the synaptic cleft. The time required for the diffusion and the time taken to release ACh contributes to the **synaptic delay**.
6. Ligand-gated Na^+ channels on the postsynaptic membrane are regulated by the attachment and removal of ACh (ii) through its nicotinic acetylchold receptors (NAChR). They are closed when ACh is not present. If ACh is attached to the channel the gate remains opened until ACh is removed or digested.
7. Opening of the ligand-gated channels allows influx of Na^+ ions into the intracellular fluid of the postsynaptic muscle cell, creating an **endplate potential** (depolarization of the membrane at the postsynaptic junction). The end-plate potential brings the muscle membrane potential to the threshold for firing a muscle action potential. It is a graded response (unlike action potential), the more NAChRs are activated, the bigger the endplate potential produced travels across the muscle surface and triggers muscle contraction.
8. **Acetylcholinesterase**, which is weakly associated with the postsynaptic membrane of the synaptic cleft, removes ACh via hydrolysis to **acetate** and **choline**.
9. Active reuptake of choline, which is then recycled, takes place.

Answers
25. F T T F T
26. 2 – D, 3 – E, 4 – A, 6 – B, 7 – F, 8 – C

27. Case study

A 63-year-old woman was admitted to the Accident and Emergency department with a fractured distal end of the radius. After treatment she was referred to see a physician. She complained that she regularly experiences bone pain. She has noticed some loss of height and the development of a hump on her back. She has difficulty in walking as she had previously suffered from a fracture of her hip. She does not do any exercise and has never been on hormone replacement therapy since her menopause. She claims that she has a balanced diet but admits that she does not drink any milk at all as it 'gives her a bad tummy'.

A. What causes the symptoms in this patient?

B. What are the risk factors for this disease?

28. Case study

A six-year-old boy was brought to the GP by his father, who has been working abroad for two years and has just returned home. He is concerned about his son as he noticed an outward protrusion (pectus carinatum) of the sternum and a 'bowing' appearance of the legs (curvature of the tibia and femur on both the lower extremities).

A. What is the likely diagnosis?

B. What is the likely cause in this patient?

29. Case study

A 40-year-old woman presented to her GP with symmetrical joint stiffness and tenderness in her hands and wrists. The problem had started about five months ago and is worse in the morning. On examination some of her fingers are deviated to the ulnar side. Fusiform swelling, redness, and warmth of the proximal interphalangeal joint were also noted. Subcutaneous nodules were seen on the extensor surfaces of the elbows. An X-ray was taken of her hands and wrists, which showed osteoporosis at the bony articulation and also some bone erosions. Some joint effusions were also noted. What is the likely diagnosis of this patient?

30. Case study

A lady was presented to her GP with weakness of her facial muscles, sometimes so severe that she finds it difficult to open her eyes. She said the weakness gets worse later in the day. The weakness gets transiently better if she lies down to get some rest.

A. What is the likely diagnosis?

B. What treatment is available for the patient?

ACh, acetylcholine; GP, general practitioner; Ca^{2+}, calcium ion

EXPLANATION: CLINICAL SCENARIOS

27. This patient suffers from **Colles fracture** (fracture of distal radius when she fell with an outstretched hand) secondary to **osteoporosis**. Osteoporosis is characterized by a decreased bone mass (**osteopenia**). This results in thin and fragile bones, which are susceptible to fracture. Compression fracture of the vertebrae may also occur (hence loss of height in the patient). It is common in postmenopausal women and in the elderly. Other risk factors include **oestrogen deficiency**, **lack of exercise**, **malnutrition** (Ca^{2+}, vitamin D, vitamin C or protein deficiencies), **immobilization**, some **endocrine diseases**, patients on **corticosteroids**, and some **genetic diseases**.

28. The patient is likely to be suffering from **rickets**. This disease is characterized by a decreased mineralization of newly formed bone due to a deficiency or abnormal metabolism of **vitamin D**. In this patient the likely cause of rickets is **malnutrition**. Rickets occurs in children prior to the closure of epiphysis. Both the remodelled bone and new bone formed at the epiphyseal growth plate are undermineralized. Endochondral bone formation is affected, and skeletal deformities result.

29. This patient has the characteristics of **rheumatoid arthritis**. It is a systemic inflammatory disease characterized by progressive arthritis, production of rheumatoid factor detectable in the blood, and patients are usually female. It is a chronic disease, thought to be triggered by an autoimmune reaction. The disease starts off with a diffused proliferative inflammation of the synovial membrane. Proliferation of the synovium and granulation tissue occurs over the articular cartilage of the joint. The joint is then fused together with fibrous tissues, leading to skeletal deformities.

30. This lady is likely to have **myasthenia gravis**. It is an autoimmune disease, characterized by neuromuscular weakness caused by the presence of autoantibodies against the neuromuscular junction. Patients are usually female, complaining of muscular weakness especially the facial muscles. Weakness of the eyelids and muscles of the eyes may lead to drooping eyelids (ptosis) and double vision (diplopia). The weakness is characteristically worsened with repeated contractions, and gets worse later in the day. Respiratory muscle involvement may lead to death. The disease can be treated by **anticholinesterase agents**. They decrease the breakdown rate of ACh, and enhance the competitiveness of ACh against the antibiotics on the nicotinic receptors on the postsynaptic membrane.

Answers

27. See explanation
28. See explanation
29. See explanation
30. See explanation

EXPLANATION: SKELETAL MUSCLES AND MUSCLE CONTRACTION AT MOLECULAR LEVEL (ii) THE CROSS-BRIDGE CYCLE

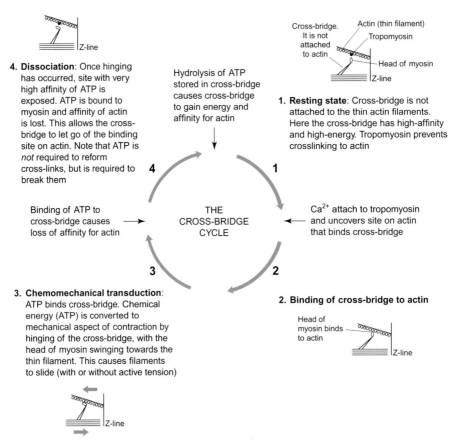

4. Dissociation: Once hinging has occurred, site with very high affinity of ATP is exposed. ATP is bound to myosin and affinity of actin is lost. This allows the cross-bridge to let go of the binding site on actin. Note that ATP is *not* required to reform cross-links, but is required to break them

Binding of ATP to cross-bridge causes loss of affinity for actin →

3. Chemomechanical transduction: ATP binds cross-bridge. Chemical energy (ATP) is converted to mechanical aspect of contraction by hinging of the cross-bridge, with the head of myosin swinging towards the thin filament. This causes filaments to slide (with or without active tension)

Hydrolysis of ATP stored in cross-bridge causes cross-bridge to gain energy and affinity for actin

Cross-bridge. It is not attached to actin — Actin (thin filament) / Tropomyosin
Head of myosin
Z-line

1. Resting state: Cross-bridge is not attached to the thin actin filaments. Here the cross-bridge has high-affinity and high-energy. Tropomyosin prevents crosslinking to actin

THE CROSS-BRIDGE CYCLE

Ca^{2+} attach to tropomyosin ← and uncovers site on actin that binds cross-bridge

2. Binding of cross-bridge to actin

Head of myosin binds to actin
Z-line

The **cross-bridge interaction** is the interaction between actin and myosin filaments in a muscle cell. It is a **chemomechanical transduction** process, during which chemical energy of ATP is transformed into mechanical energy of muscle cells. **Contraction** is the continuous cycle of such cross-bridge interactions (the **cross-bridge cycle**).
Every time a cross-bridge completes a single cycle, one ATP is hydrolysed to provide energy for mechanical contraction of the muscle.
Cross-bridge cycling (muscle contraction) continues until either Ca^{2+} is withdrawn (normal) or ATP is depleted (pathological).

ATP, adenosine triphosphate; Ca^{2+}, calcium ion

SECTION 2

THE HEAD AND NECK

2 THE HEAD AND NECK

1. With regard to the cranium

a. The pterion is where the sutures of the frontal, parietal, temporal and sphenoid bones all meet
b. The pterion resembles the letter 'H'
c. The sagittal suture is the meeting point of the occipital bone and the two parietal bones
d. The coronal suture is where the frontal and two parietal bones meet
e. The lambdoid suture separates the occipital and two parietal bones

2. Concerning the base of the cranium

a. The parietal bones form part of the base of the cranium
b. The spinal cord passes through the foramen magnum in the occipital bone
c. The base of the cranium is divided into anterior, middle and posterior fossae
d. The mastoid process is part of the occipital bone
e. The styloid process is part of the temporal bone

3. True or false? With regard to the face

a. The maxilla is the only bone in the skull not to be connected via immovable joints
b. The mandible articulates with the temporal bone via a synovial joint that has both gliding and hinge-type properties
c. The ethmoid bone forms part of the orbit
d. The glabella is part of the frontal bone
e. The zygomatic arch consists of the zygomatic bone

4. Match the foramina below with the structures that pass through them

Options

A. Internal jugular vein
B. Hypoglossal nerve
C. Olfactory nerve
D. Facial nerve
E. Spinal cord
F. Facial nerve and hypoglossal nerve
G. Motor component of facial nerve
H. Optic nerve
I. Internal carotid artery
J. No structure

1. Foramen magnum
2. Carotid canal
3. Jugular foramen
4. Foramen lacerum
5. Stylomastoid foramen

CN, cranial nerve

EXPLANATION: THE CRANIUM AND THE FACE

The skull can be divided into two parts: the **cranium** and the **face**. The **cranium** is composed of **eight bones**: one frontal, two parietal, two temporal, one occipital, the sphenoid and the ethmoid.

The **pterion** is the anatomical landmark where the sutures of the frontal, parietal, temporal and sphenoid bones all meet. It **resembles the letter 'H'**. The two parietal bones are separated by the **sagittal suture**. The frontal bone meets both parietal bones at the **coronal suture**. The occipital bone meets the two parietal bones at the **lambdoid suture**.

The **base of the cranium** is divided into anterior, middle and posterior fossae. It is formed by the frontal, two temporal, occipital, ethmoid and sphenoid bones. The **foramen magnum** is the largest of the foramina in the base of the cranium, through it passes the **spinal cord**. Looking from below, as well as all the foramina, a pair of prominences are apparent on the temporal bones: the **mastoid processes** and the **styloid processes**.

The **face** is made up of **ten bones**: two nasal, two vomer (inferior conchae), two zygomatic, two lacrimal, the maxilla, the **mandible** (which articulates via a **synovial** joint with gliding and hinge-type properties) and the superior and middle conchae (not regarded as separate bones since they are projections of the ethmoid into the nasal cavity). The frontal and sphenoid bones also form part of the face. The glabella is the part of the frontal bone between the two eyebrows.

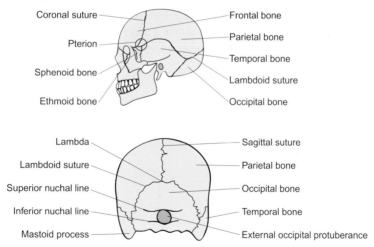

See page 34 for further diagrams.

The **orbit** is formed by the frontal, ethmoid, sphenoid, lacrimal, maxillary and zygomatic bones. The **zygomatic arch** is the arc of bone on either side of the face below the eyes. It consists of connected processes from both the zygomatic and temporal bones.

Answers

1. T T F T T
2. F T T F T
3. F T T T F
4. 1 – E, 2 – I, 3 – A, 4 – J, 5 – G

5. Use the options below to label the diagram of the temporomandibular joint

Options

A. Lateral pterygoid muscle
B. Articular eminence
C. Articular fossa
D. Condylar process of mandible
E. Cut-away of zygoma
F. Joint capsule
G. Articular disc
H. Upper and lower compartments

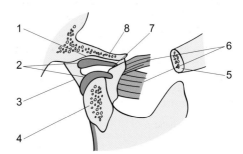

6. True or false? In the neonate's skull

a. The mandibular symphysis is still open
b. The fontanelles are usually closed by the first year
c. There is no mastoid process until the second year
d. The frontal (metopic) suture closes during the sixth year
e. The external acoustic meatus and position of the tympanic membrane resemble those in the adult skull

---

EXPLANATION: THE BASE OF THE SKULL

The temporomandibular joint is formed by the articulation between the condylar process of the mandible and the mandibular fossa of the temporal bone. It is a **synovial joint** that has both gliding and hinge-type properties. The joint cavity is divided into upper and lower compartments by a disc of dense fibrous connective tissue (**articular disc/meniscus**), which is attached to the capsule of the joint. The lower chamber facilitates hinge-like movements (elevation and depression), while the upper chamber allows gliding movements. The muscles of mastication are shown in the table below.

Muscle	Origin	Insertion	Action
Masseter	Zygomatic arch	Lateral surface of ramus of mandible	Elevates and protrudes mandible
Temporalis	Temporal fossa floor	Condylar process of mandible	Superior & anterior fibres elevate the mandible; posterior fibres retract the mandible
Medial pterygoid	Tuberosity of maxilla & lateral pterygoid plate	Medial surface of angle of mandible	Raises mandible
Lateral pterygoid	Greater wing of sphenoid & lateral pterygoid plate	Anterior surface of condylar process of mandible	Opens the jaw, grinding action side to side, protrusion

In the neonate the **mandibular symphysis** and **fontanelles close** by the **second year**. The **external acoustic meatus** is **shorter** and the **tympanic membrane is closer to the surface** of the skull. The **frontal (metopic) suture closes** during the **sixth year**. The **mastoid process** forms during the **second year**. The structures of the neonate skull are shown in the figures below.

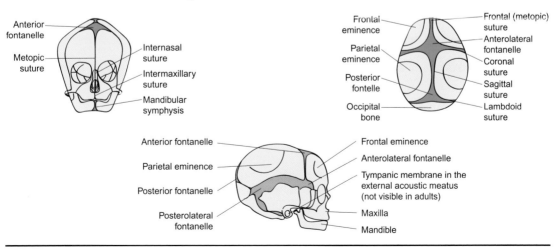

Answers

5. 1 – C, 2 – H, 3 – F, 4 – D, 5 – E, 6 – A, 7 – G, 8 – B
6. T F T T F

7. List the five layers of the scalp

8. Consider the muscles of the skull

    a. The sternocleidomastoid is the main extensor of the head
    b. The temporalis passes over the zygomatic arch
    c. The temporalis is supplied by the facial nerve
    d. The temporalis and masseter both close the jaw
    e. The masseter is supplied by the facial nerve

9. With regard to the muscles of facial expression

    a. The orbicularis oculi is supplied by the facial nerve which stimulates it to close the eye
    b. The frontalis elevates the eyebrows
    c. The orbicularis oris is supplied by the facial nerve
    d. The levator labii superioris dilates the nostrils
    e. The nasalis dilates the nostrils

EXPLANATION: THE SCALP AND MUSCULATURE

The layers of the **scalp** are:

1. **Skin**
2. **Connective tissue**
3. **Aponeurosis**
4. **Loose areolar tissue**
5. **Pericranium (periostium)** (7)

Muscles of the skull may be subdivided into two groups: the **great muscles of the skull** and the **muscles of facial expression**.

The **great muscles of the skull** include the stern-ocleidomastoid, temporalis and masseter. The temporalis (which passes under the zygomatic arch) and masseter are supplied by the **trigeminal nerve**. They both **close the jaw**. The sternocleidomastoid is supplied by the **spinal accessory nerve** and is the **main extensor of the head**.

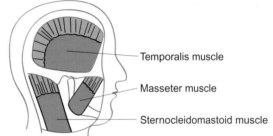

Temporalis muscle

Masseter muscle

Sternocleidomastoid muscle

Muscles of facial expression include the frontalis, orbicularis oculi, nasalis, levator labii superioris, levator anguli oris, orbicularis oris, buccinator, depressor labii inferioris, depressor anguli oris, mentalis and platysma. They are all supplied by the **facial nerve**.

The orbicularis oculi only closes eyelids. The frontalis elevates the eyebrows.

The orbicularis oris brings the lips together and protrudes them. The levator labii superioris elevates the upper lip as well as dilating the nostrils. The nasalis draws the sides of the nose medially. The depressor labii inferioris lowers the lower lip while

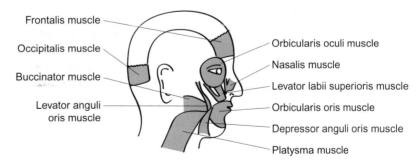

Frontalis muscle

Occipitalis muscle

Buccinator muscle

Levator anguli oris muscle

Orbicularis oculi muscle

Nasalis muscle

Levator labii superioris muscle

Orbicularis oris muscle

Depressor anguli oris muscle

Platysma muscle

the mentalis protrudes it. The levator anguli oris raises the angle of the mouth while the depressor anguli oris lowers it. The buccinator draws the cheeks towards the teeth. The platysma tenses the skin of the lower face and neck.

Answers

7. See explanation
8. T F F T F
9. T T T T F

10. Indicate whether the following statements concerning the tongue are true or false

    a. The prime function of the extrinsic muscles is to alter the shape of the tongue
    b. All of the muscles of the tongue are innervated by the hypoglossal nerve (CN XII)
    c. The intrinsic muscles form the body of the tongue
    d. The longitudinal, transverse and vertical muscle fibres constitute the intrinsic muscles of the tongue
    e. In unilateral hypoglossus nerve damage the protruding tongue deviates to the opposite side of that bearing the lesion

11. Using your knowledge of the extrinsic muscles of the tongue, fill in the gaps in the table below

Muscle	Origin	Insertion	Action
Hypoglossus	Greater horn of hyoid bone	Merges with styloglossus and genioglossus muscles	
	Genial spine of mandible	Forms bulk of tongue	
Palatoglossus	Palatine aponeurosis		Pulls root of tongue upward and backwards
Styloglossus	Styloid process	Merges with hyoglossus and genioglossus muscles	

12. Label the diagram below with the nerves that are responsible for both general sensation and chemoreception of the tongue

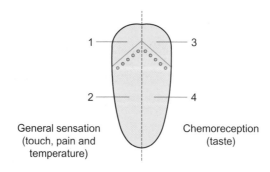

General sensation
(touch, pain and
temperature)

Chemoreception
(taste)

CN, cranial nerve

EXPLANATION: THE MUSCLES OF THE TONGUE

The muscles of the tongue are categorized into **extrinsic** (originating outside the tongue) and **intrinsic** (forming the substance of the tongue). The extrinsic muscles are the **genioglossus, hyoglossus, palatoglossus** and **styloglossus**. The longitudinal, transverse and vertical muscle fibres constitute the intrinsic muscles of the tongue. All the muscles of the tongue are innervated by the **hypoglossal nerve** (CN XII), with the exception of the palatoglossus which is supplied by the pharyngeal plexus. In unilateral hypoglossus nerve damage the protruding tongue deviates to the side bearing the lesion.

The dorsum of the tongue is divided by a V-shaped line, known as the **sulcus terminalis**, into an anterior two-thirds and a posterior third. The **anterior two-thirds** is drained by submental lymph nodes and gains sensory innervation from the lingual nerve (CN V$_3$) (general sensation) and chorda tympani (CN VII) (chemoreception). The **posterior third** is drained by deep cervical nodes, and sensory innervation (for both chemoreception and general sensation) is derived from the glossopharyngeal nerve (CN IX). The lingual artery, tonsillar branch of the facial artery and ascending pharyngeal artery perfuse the tongue.

Muscle	Origin	Insertion	Action
Genioglossus	Genial spine of mandible	Forms bulk of tongue	Protrusion (sticking tongue out)
Hypoglossus	Greater horn of hyoid bone	Merges with styloglossus and genioglossus muscles	Depresses tongue
Styloglossus	Styloid process	Merges with hyoglossus and genioglossus muscles	Draws tongue upwards and backwards
Palatoglossus	Palatine aponeurosis	Side of the tongue	Pulls root of tongue upward and backwards

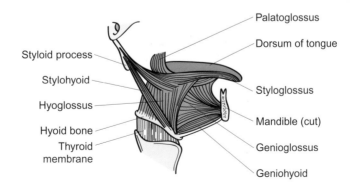

Palatoglossus

Dorsum of tongue

Styloid process

Stylohyoid

Styloglossus

Hyoglossus

Mandible (cut)

Hyoid bone

Thyroid membrane

Genioglossus

Geniohyoid

Answers

10. F F T T F
11. See table
12. 1 – glossopharyngeal nerve (CN IX), 2 – lingual nerve (CN V$_3$), 3 – glossopharyngeal nerve (CN IX), 4 – chorda tympani (CN VII)

13. Concerning the anatomy of the orbits

    a. The inferior wall is only formed by the orbital plate of the maxilla and no other bones
    b. The medial wall is only composed of the ethmoid and lacrimal bones
    c. The superior wall is formed by the frontal bone only
    d. One of the components of the lateral wall is the greater wing of the sphenoid
    e. The superior orbital fissure communicates with the pterygopalatine fossa

14. Consider the relationship between the cranial nerves and eye movements. Which cranial nerve is involved in the following movements of the right eye?

1  2   3   4 5

15. Using the options provided, label the diagram below

Options

    A. Lateral rectus muscle   **B.** Inferior rectus muscle
    C. Medial rectus muscle   **D.** Superior oblique muscle
    E. Inferior oblique muscle   **F.** Superior rectus muscle

CN, cranial nerve

EXPLANATION: THE ORBIT AND EXTRAOCULAR MUSCLES

Each orbit is composed of four orbital walls. The **superior wall** is formed by the frontal bone and lesser wing of the sphenoid bone. The **inferior wall** is formed by the orbital plate of the maxilla. The **lateral wall** consists of the zygomatic bone and greater wing of the sphenoid bone, and finally the **medial wall** which is formed by the orbital lamina of the ethmoid and lacrimal bones.

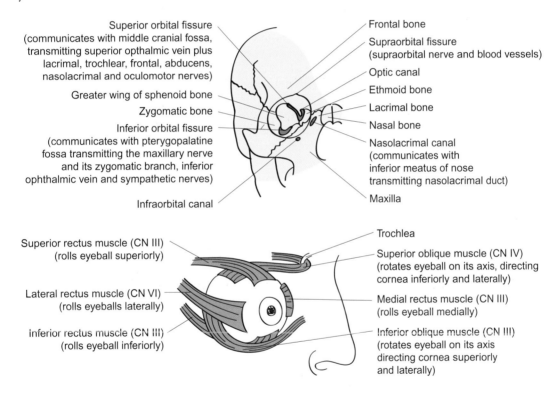

Superior orbital fissure
(communicates with middle cranial fossa,
transmitting superior opthalmic vein plus
lacrimal, trochlear, frontal, abducens,
nasolacrimal and oculomotor nerves)

Greater wing of sphenoid bone

Zygomatic bone

Inferior orbital fissure
(communicates with pterygopalatine
fossa transmitting the maxillary nerve
and its zygomatic branch, inferior
ophthalmic vein and sympathetic nerves)

Infraorbital canal

Frontal bone

Supraorbital fissure
(supraorbital nerve and blood vessels)

Optic canal

Ethmoid bone

Lacrimal bone

Nasal bone

Nasolacrimal canal
(communicates with
inferior meatus of nose
transmitting nasolacrimal duct)

Maxilla

Superior rectus muscle (CN III)
(rolls eyeball superiorly)

Lateral rectus muscle (CN VI)
(rolls eyeballs laterally)

Inferior rectus muscle (CN III)
(rolls eyeball inferiorly)

Trochlea

Superior oblique muscle (CN IV)
(rotates eyeball on its axis, directing
cornea inferiorly and laterally)

Medial rectus muscle (CN III)
(rolls eyeball medially)

Inferior oblique muscle (CN III)
(rotates eyeball on its axis
directing cornea superiorly
and laterally)

Eye movements are controlled by six extraocular muscles. The superior oblique muscle is innervated by the **fourth cranial nerve (trochlear)** and the lateral rectus muscle by the **sixth (abducens nerve).** All other extraocular muscles gain their innervation from the **third cranial nerve (oculomotor).**

Answers

13. T T F T F
14. 1 – CN VI, 2 – CN IV, 3 – CN III, 4 – CN III, 5 – CN III
15. 1 – A, 2 – B, 3 – E, 4 – C, 5 – D, 6 – F

16. With regard to the blood supply of the head and neck

a. The common carotid and vertebral arteries provide the main blood supply of the head and neck

b. The maxilla is supplied by a branch of the internal carotid artery

c. The face is supplied by the facial artery – a branch of the external carotid artery

d. The scalp is supplied by three branches of the external carotid artery: the superficial temporal, posterior auricular and occipital arteries

e. The orbit is supplied by the ophthalmic artery – a branch of the internal carotid artery

17. Consider the blood supply of the head and neck

a. Drainage of the head and neck is via the internal and external jugular veins

b. Deep and superficial venous systems of the head and neck do not communicate

c. The drainage of the brain is via venous sinuses and plexuses into the internal and external jugular veins

d. The external jugular vein has three main branches: the facial, superficial temporal and posterior auricular veins

e. The internal jugular vein receives veins corresponding to the branches of the external carotid artery

18. Consider dermatomes of the head and neck

a. The sensory supply of the head is provided by C1 and C2

b. The trigeminal nerve supplies most of the face and scalp

c. C2 provides the sensory supply of the neck

d. The sensory innervation of the face is provided by only two divisions of the trigeminal nerve

e. The angles of the mandible and the ears are innervated by C2

CN, cranial nerve

EXPLANATION: VASCULATURE OF THE HEAD AND NECK

The **common carotid and vertebral arteries** provide the main blood supply of the head and neck. The **common carotid** divides into the **external and internal carotid** arteries. The **external** carotid artery **mainly** supplies the **face and scalp** while the **internal** carotid **and vertebral** arteries supply the **cerebral hemispheres, cerebellum** and **brainstem**. The **external carotid** artery gives off **eight branches,** which are shown below in a side view from the left in relation to the mandible.

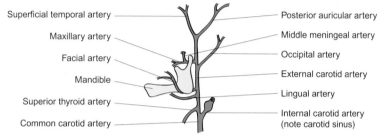

1. **Superior thyroid artery** – supplies the thyroid gland
2. **Ascending pharyngeal artery**
3. **Lingual artery** – supplies the floor of the mouth
4. **Facial artery** – supplies the face
5. **Maxillary artery** – supplies the maxilla
6. **Occipital artery** – supplies the scalp
7. **Superficial temporal artery** – supplies the scalp and forehead
8. **Posterior auricular artery** – supplies the scalp

The **middle meningeal** artery is a branch of the maxillary artery. It supplies the dura mater and the bones of the cranium.

The **internal carotid** artery has **four branches:** the **ophthalmic** artery which supplies the eye and muscles of eye movements, and the **anterior cerebral, middle cerebral** and **posterior communicating** arteries which supply the cerebral hemispheres.

Blood drains from the brain via veins into the **dural venous sinuses**. These sinuses **drain into the internal jugular veins**. The latter also receive veins corresponding to the branches of the external carotid arteries. The **internal jugular veins** therefore **drain the contents of the skull, the face, scalp and neck**. The **external jugular veins** drain the **neck and** the **posterior aspect of** the **scalp**.

Deep and superficial veins of the head and neck communicate, and by doing so allow spread of infections from the face to the meninges and/or the brain.

The sensory supply of the **face and scalp** is mainly through the **three divisions** of the **trigeminal nerve** (ophthalmic (CN V$_1$), maxillary (CN V$_2$) and mandibular (CN V$_3$)). **C3** supplies the **neck** while **C2** supplies the **angles of** the **mandible and** the **pinnas** of the ears. C1 has no cutaneous distribution.

Answers
16. T F T T T
17. T F T T T
18. F T F F T

19. True or false? Bell's palsy

    a. Occurs unilaterally

    b. Involves paralysis of the trigeminal nerve

    c. Predominately affects those aged between 30 and 50 years

    d. May be preceded by hyperacusis

    e. Resolves in most patients three months after the initial onset of symptoms

20. Which of the following is the most unlikely consequence of a blunt blow to the left side of the face at the level of the eye?

    a. Splinters of bone impregnating themselves onto the surface of the left eye

    b. The extra-ocular muscles becoming trapped disturbing free movement of the left eye

    c. Damage to the maxillary division of the left trigeminal nerve

    d. Tracking of infections from the nasal sinuses into the left orbit

    e. Bilateral temporomandibular joint dislocation

CN, cranial nerve

EXPLANATION: CLINICAL SCENARIOS

19. Bell's palsy is the unilateral paralysis of the facial muscles which occurs suddenly. A widely held view is that it is the result of a viral neuropathy that causes inflammation of the facial nerve (CN VII). Although it may occur in anyone, it is most common in those **aged between 30 and 50** years of age. About two days before the condition strikes, patients experience pain behind the ear, on the side to be affected, or an extreme sensitivity to sounds (**hyperacusis**). The condition leaves the patient with a drooping smile, difficulty articulating speech (**dysarthria**), and on the affected side an inability to raise the eyebrow and a very watery eye due to the turning-out of the lower lid. The paralysis worsens over two days and recovery begins after two weeks, with **most patients making a full recovery three months after the initial onset of symptoms.**

Eye • When lid closed eye rolls upward
     • Watery eye, as there is failure of eye
       closure (not ptosis)

Face • Unilateral facial paralysis

Mouth • Dysarthria
      • Difficulty chewing
      • Taste impairment
      • Drooling

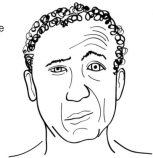

20. Traumatic injuries to the upper face damage the thin bones most readily. If it is a **symmetrical** blunt blow to the nose the nasal bones are easily broken. However, an **asymmetrical** blow at the level of the eye can also fracture the **lacrimal** and **ethmoid** bones that form the medial wall of an orbit. In the case of a lacrimal bone fracture, tears may overflow the lower lid as the lacrimal sac, which carries tears away from the eye, may be disrupted. As the ethmoid bones separate the eyes from the nasal sinuses medially, a fracture here may result in nasal bacteria tracking back into the orbit and then to the cavernous sinus. It is likely that such an infection would be life-threatening.

In a situation where there is a hefty blow to the cheek, the medial orbital floor is at great risk of damage; such damage is referred to as a **'blow-out fracture'**. This type of breakage may be complicated by splinters of bone impregnating themselves onto the surface of the eye; the extra-ocular muscles becoming trapped disturbing free movement of the eye; damage to the maxillary division (CN V_2) of the trigeminal nerve (CN V) as well as tracking of infections from the nasal sinuses to the cavernous sinus, as described above.

Answers
19. T F T T T
20. F F F F T

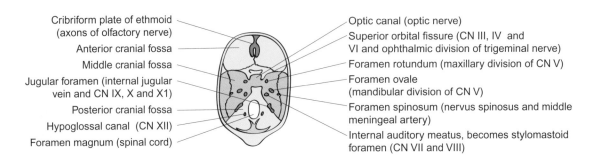

Cribriform plate of ethmoid
(axons of olfactory nerve)

Anterior cranial fossa

Middle cranial fossa

Jugular foramen (internal jugular
vein and CN IX, X and X1)

Posterior cranial fossa

Hypoglossal canal (CN XII)

Foramen magnum (spinal cord)

Optic canal (optic nerve)

Superior orbital fissure (CN III, IV and
VI and ophthalmic division of trigeminal nerve)

Foramen rotundum (maxillary division of CN V)

Foramen ovale
(mandibular division of CN V)

Foramen spinosum (nervus spinosus and middle
meningeal artery)

Internal auditory meatus, becomes stylomastoid
foramen (CN VII and VIII)

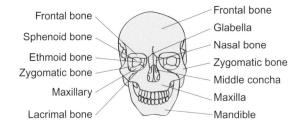

Frontal bone

Sphenoid bone

Ethmoid bone

Zygomatic bone

Maxillary

Lacrimal bone

Frontal bone

Glabella

Nasal bone

Zygomatic bone

Middle concha

Maxilla

Mandible

SECTION 3

THE TRUNK

1. Concerning the thoracic cage

a. Ribs 8–12 are termed false ribs as they indirectly articulate with the sternum
b. The first seven pairs of ribs directly articulate with the sternum and are known as the true ribs
c. A typical rib has a head, neck, tubercle, angle and shaft
d. Ribs 1, 2, 11 and 12 are considered to be atypical
e. From superior to inferior, neurovascular structures run within the subcostal groove as nerve, artery and vein

2. Examine the seventh rib and label it from the options below

Options

A. Articular part of tubercle
B. Neck
C. Tubercle
D. Demifacet for vertebral bodies of T6 and T7
E. Shaft
F. Head

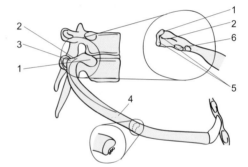

3. Use the options below to label the diagram of the thoracic ribcage

Options

A. Costal cartilage
B. First thoracic vertebra
C. Body
D. Manubrium
E. Xiphoid process
F. First rib
G. Twelfth rib
H. Second rib
I. Sternal angle (of Louis)

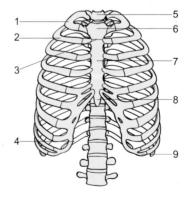

EXPLANATION: THE THORACIC CAGE

The thoracic cage includes the **12 pairs of ribs** articulating posteriorly with the **vertebral column** and the **sternum**. The **sternum** is divided into three parts: the **manubrium (superiorly)**, the **body** and the **xiphoid process (inferiorly)**. **True ribs** are the first seven as they are attached directly to the sternum by their own costal cartilages. **False ribs** (8–10 inclusive) indirectly articulate with the sternum, as their costal cartilage is combined with that of the rib above. Ribs 11 and 12 do not articulate with the sternum and hence are termed **floating ribs**.

Typical ribs (3–10) consist of a **head, neck, tubercle, angle** and **shaft**. (Note that the tenth rib is sometimes considered as atypical).

- The **head** is wedge-shaped, with two demifacets articulating with the numerically corresponding vertebral body and the vertebra immediately above it.
- The **neck** is a flattened portion that lies lateral to the head. It separates the head from the tubercle.
- The **tubercle** is a projection that has an articulating facet that attaches to the transverse process of the numerically corresponding vertebrae.
- The **angle** lies a short distance anterior to the tubercle and is the point of greatest curvature.
- The **shaft** is the largest part of the rib that has a large, rounded superior border and thin inferior border. Behind its thinnest border there is the subcostal groove, housing an intercostal vein (superiorly), intercostal artery and intercostal nerve (remembered by the mnemonic **VAN**).

The first, second, eleventh and twelfth pairs of ribs are considered to be **atypical**.

- The first rib is the broadest and shortest of the ribs. It has a prominent scalene tubercle on the inner border of its superior surface allowing attachment of the **scalenus anterior muscle**.
- The second rib is less curved and slightly thinner than the first rib, with a broad and rough tuberosity facilitating attachment for the **serratus posterior muscle**.

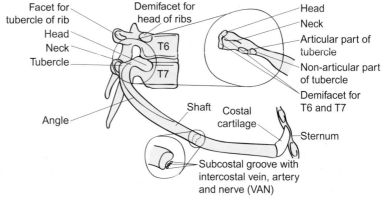

- Ribs 10, 11 and 12 only have a **single facet** upon their heads limiting them to articulation solely with their corresponding vertebral body.

Answers

1. F T T T F
2. 1 – F, 2 – B, 3 – C, 4 – E, 5 – D, 6 –A
3. 1 – F, 2 – H, 3 – I, 4 – G, 5 – B, 6 – D, 7 – C, 8 – E, 9 – A

4. The diaphragm

   a. Mainly increases the anterior–posterior dimension of the thoracic cavity during inspiration

   b. Is perfused by the musculophrenic and inferior gastric artery

   c. Receives motor fibres from the vagus and phrenic nerves

   d. Originates from the posterior surface of the xiphoid process, the lower eight pairs of ribs and the lumbar vertebra to the level of L3

   e. Consists of muscular fibres that insert upon the central tendon

5. Label this cross-sectional diagram of the thoracic wall from the options given

Options

   A. External intercostal muscle
   B. Internal intercostal muscle
   C. Rib
   D. Intercostal artery
   E. Intercostal nerve
   F. Innermost intercostal muscle
   G. Subcostal groove
   H. Intercostal vein
   I. Neurovascular bundle

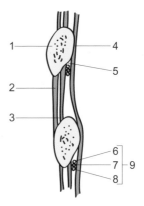

6. Label this diagram, which depicts the inferior surface of the diaphragm, using the options given

Options

   A. Inferior vena cava
   B. Lumbar vertebra
   C. Psoas major muscle
   D. Oesophagus
   E. Central tendon
   F. Ribs
   G. Aorta
   H. Xiphoid process
   I. Quadratus lumborum muscle

EXPLANATION: MUSCLES OF RESPIRATION

The intercostal muscles are arranged as three muscular layers innervated by the anterior rami of the first 11 pairs of thoracic spinal nerves (T2–T12).

External intercostal muscles originate from the inferior border of the rib above, inserting upon the superior border of the rib below. They form the **outermost muscular layer** with fibres **running forward** and **downwards** (think 'hands in pockets').

Internal intercostal muscles form the middle layer. They also originate from the inferior border of the rib above, inserting upon the superior border of the rib below. However, **fibres** run at **90 degrees** to those of the external intercostals in a **downward** and **backwards** direction.

The **innermost intercostals** are separated from the internal intercostals by nerves and vessels. These muscles form the **deepest layer** and include **subcostal** (lateral and posterior parts of thoracic cage) and **transversus thoracic muscles** (anterior part of thoracic cage).

The diaphragm is a bi-domed-shaped muscle innervated by the **phrenic nerve (C3, C4** and **C5)** and perfused by the **musculophrenic** and **inferior phrenic arteries**. It is the most important muscle involved in respiration. As it contracts, the floor of the thoracic cavity is drawn downward, increasing the superior–inferior dimension of the cavity. The diaphragm's muscular fibres originate from the posterior surface of the **xiphoid process**, **lower six pairs of ribs** and the **lumbar vertebra** to the level of **L3**. All of these fibres then insert upon a strong aponeurosis known as the **central tendon**. Structures traverse the diaphragm through the aortic opening, oesophageal opening or the vena caval opening.

See page 54 for a table that gives the muscles involved in respiration.

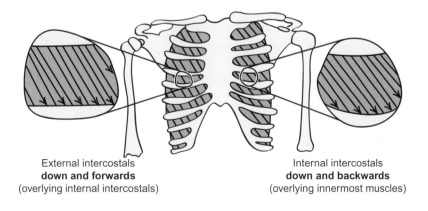

External intercostals
down and forwards
(overlying internal intercostals)

Internal intercostals
down and backwards
(overlying innermost muscles)

Answers
4. F F F F T
5. 1 – C, 2 – A, 3 – B, 4 – F, 5 – G, 6 – H, 7 – D, 8 – E, 9 – I
6. 1 – H, 2 – F, 3 – A, 4 – D, 5 – G. 6 – I, 7 – C, 8 – B, 9 – E

7. Make short notes on the arrangement of the abdominal wall muscles

8. Label the diagram below using the list provided

Options

A. Transversus abdominis muscle
B. Rectus abdominis muscle
C. Pectoralis major muscle
D. External oblique muscle
E. Tendinous intersection
F. Linea alba
G. Internal oblique muscle
H. Serratus anterior muscle

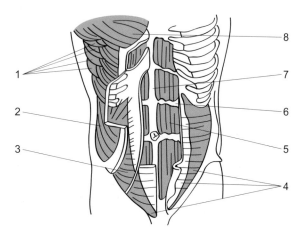

9. Label the diagram below which is a transverse section through the rectus sheath above the level of the umbilicus

Options

A. Superior epigastric artery
B. Rectus sheath
C. Internal oblique muscle
D. Skin
E. Transversus abdominis muscle
F. Rectus abdominis muscle
G. External oblique muscle
H. Linea alba

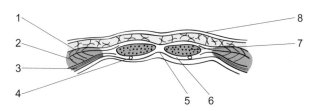

EXPLANATION: ANTERIOR ABDOMINAL WALL MUSCLES

From superficial to deep, the lateral abdominal muscles are the **external oblique**, **internal oblique** and **transversus abdominis**. They mimic the muscular arrangement observed in the thoracic wall, but form aponeuroses towards the anterior part of the abdomen.

Either side of the midline, is the **rectus abdominis** muscle extending from the xiphoid process of the sternum and fifth to seventh costal cartilages, to the pubic symphysis and pubic crest below. The aponeuroses enclose the rectus abdominis, forming the **rectus sheath**; fusing at the midline (**linea alba**) and lateral edge (**linea semilunaris**).

Fibres of the **external oblique** muscle arise from the lower eight ribs and run downwards and forwards to form an aponeurosis **anteriorly** that lies over the rectus abdominis. Inferiorly the aponeurosis stretches over the pubic tubercle and inserts onto the anterior superior iliac spine; then folding back upon itself to form the **inguinal ligament**. The fibres of the **internal oblique** muscle run upwards and forwards arising from the thoracolumbar fascia, iliac crest and lateral two-thirds of the inguinal ligament. Like the external oblique muscle, they become aponeurotic **anteriorly**. The inferior part of the aponeurosis inserts into the pubic crest, fusing with the aponeurosis of the transversus abdominis muscle to form the **conjoint tendon**.

The **transversus abdominis** muscle lies deep to the internal oblique muscle and its fibres run horizontally. They originate from the thoracolumbar fascia, iliac crest and lateral third of the inguinal ligament. Superiorly, its aponeurosis passes **behind** the rectus abdominis, but inferiorly this aponeurosis lies in front of the rectus abdominis.

All of the anterior abdominal wall muscles contribute to the variety of trunk movements. They also aid in compression of the abdominal cavity during processes such as defecation. The muscles of the anterior abdominal wall, from superficial to deep (left to right) are shown below:

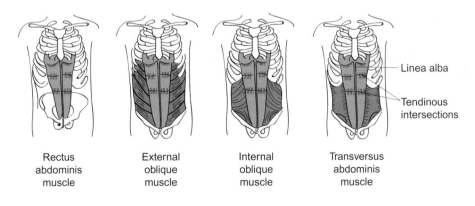

Rectus abdominis muscle · External oblique muscle · Internal oblique muscle · Transversus abdominis muscle · Linea alba · Tendinous intersections

---

Answers

7. See explanation
8. 1 – H, 2 – D, 3 – G, 4 – A, 5 – B, 6 – E, 7 – F, 8 – C
9. 1 – G, 2 – C, 3 – E, 4 – A, 5 – H, 6 – B, 7 – F, 8 – D

10. Regarding the vessels of the abdominal wall

a. The superior and inferior epigastric arteries lie anterior to the rectus abdominis muscle
b. The inferior epigastric artery is a branch of the external iliac artery
c. Superficial drainage below the umbilicus is facilitated by the femoral veins to the inferior vena cava
d. Patients with liver cirrhosis may develop dilatation of the superficial veins in this region
e. Several branches of the descending aorta contribute to the blood supply

11. Consider the nerves and arteries of the abdominal wall and label the diagram below using the options provided

Options

A. Femoral artery
C. Transversus abdominis muscle
E. Tenth thoracoabdominal nerve
G. Iliohypogastric nerve (L1)
I. Ilioinguinal nerve (L1)
K. Superficial epigastric artery
L. Seventh thoracoabdominal nerve
N. Eighth thoracoabdominal nerve
P. Ninth thoracoabdominal nerve

B. Musculophrenic artery
D. Superior epigastric artery
F. Subcostal nerve (T12)
H. Inferior epigastric artery
J. Tenth and eleventh posterior intercostal arteries
M. External iliac artery
0. Eleventh thoracoabdominal nerve
Q. Rectus abdominis muscle

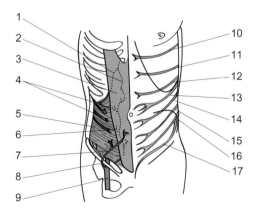

12. Write short notes on the venous drainage of the anterior abdominal wall

EXPLANATION: NERVES AND VESSELS OF THE ANTERIOR ABDOMINAL WALL

Segmental innervation of the abdominal wall is derived from the anterior rami of the **lower six intercostal nerves (T7–T12)** and the **first lumbar nerves (L1)**. Due to the location of the intercostal nerves within the abdomen, they are collectively known as the **thoracoabdominal nerves**. They innervate the anterior and lateral muscles as well as the skin of the abdomen. L1 supplies the lateral abdominal wall muscles and adjacent skin via the **ilioinguinal** and **iliohypogastric nerves**.

The **blood supply** of the abdominal wall is provided by vessels that run vertically and obliquely. Vertically, we have the **superior** and **inferior epigastric arteries** that travel within the rectus sheath behind the rectus abdominis muscle. The superior epigastric artery is a terminal branch of the **internal thoracic artery** and it penetrates the rectus sheath at the level of the xiphoid process. Travelling behind the rectus abdominis, it anastamoses freely with the inferior epigastric artery at the muscle's upper part. The inferior epigastric artery is a branch of the **external iliac artery** and reaches this point of anastamosis by extending up from behind the ductus deferens.

Obliquely, there are the **posterior arteries** of the **lower six spaces**, and the **four lumbar arteries**. These branches of the descending aorta aid the branches of the anastamosis between the epigastric arteries to supply the lateral and anterior abdominal wall.

Superficial venous drainage of the anterior abdominal wall passes above, via the **lateral thoracic vein** to the **axillary vein** and below via the **superficial epigastric** and **great saphenous veins** to the **femoral vein**. Also, veins of the umbilical region are connected to the **portal vein** via small **paraumbilical veins (12)**.

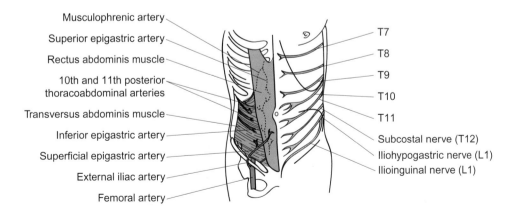

Musculophrenic artery
Superior epigastric artery
Rectus abdominis muscle
10th and 11th posterior thoracoabdominal arteries
Transversus abdominis muscle
Inferior epigastric artery
Superficial epigastric artery
External iliac artery
Femoral artery

T7
T8
T9
T10
T11
Subcostal nerve (T12)
Iliohypogastric nerve (L1)
Ilioinguinal nerve (L1)

Answers

10. F T T T T
11. 1 – B, 2 – D, 3 – Q, 4 – J, 5 – C, 6 – H, 7 – K, 8 – M, 9 – A, 10 – L, 11 – N, 12 – P, 13 – E, 14 – O, 15 – F, 16 – G, 17 – I
12. See explanation

13. Label the diagram of the hip bone below using the options provided

Options

A. Pubic symphysis
B. Posterior superior iliac spine
C. Iliac crest
D. Anterior inferior iliac spine
E. Pubic tubercle
F. Anterior superior iliac spine
G. Ischial tuberosity
H. Spine of ischium
I. Lesser sciatic notch
J. Greater sciatic notch
K. Acetabulum

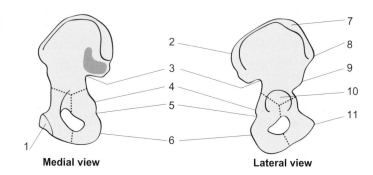

Medial view   Lateral view

14. Label the diagram below with the options provided

Options

A. Sacroiliac joint
B. Pubic tubercle
C. Acetabulum
D. Anterior superior iliac spine
E. Pubic crest
F. Anterior inferior iliac spine
G. Ischial spine
H. Obturator foramen
I. Pubic symphysis
J. Coccyx

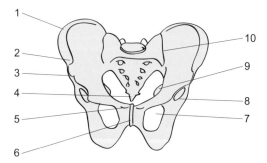

15. Concerning the pelvis

a. Define the borders of the pelvic inlet and outlet
b. Make short notes on the differences between the male and female pelvis

EXPLANATION: THE BONY PELVIS

The bony pelvis consists of **two hip bones** laterally which meet anteriorly at the **pubic symphysis** and articulate posteriorly with the **sacrum**. Also, at the tip of the sacrum is the **coccyx**. The hip bones are the result of fusion between three smaller bones at the acetabulum: the **ilium**, **ischium** and **pubis**. The **pelvic brim/inlet** is formed by the pubic symphysis anteriorly, iliopectineal lines laterally, and sacral promontory posteriorly. Below the brim is the **true pelvis** and above the brim is the **false pelvis**. At the inferior aspect of the bony pelvis, is the **pelvic outlet**. The outlet is bounded anteriorly by the pubic arch, laterally by ischial tuberosities and behind by the coccyx **(15a)**.

The ilium is the **largest** of the **three subdivisions** of the hip bone. It lies superiorly in relation to the ischium and pubis. The bony arrangement of the latter two bones forms the **obturator foramen**.

The differences between the male and female pelvis are described in the table below **(15b)**.

Feature	Male	Female
Superior pelvic aperture	Usually heart-shaped	Usually oval-shaped
Inferior pelvic aperture	Relatively narrow	Relatively wide
Infra-pubic angle	Acute	Obtuse
Greater sciatic notch	Less than 90 degrees	90 degrees or more
Pelvic cavity	Relatively small	Relatively spacious
Obturator foramen	Round	Oval

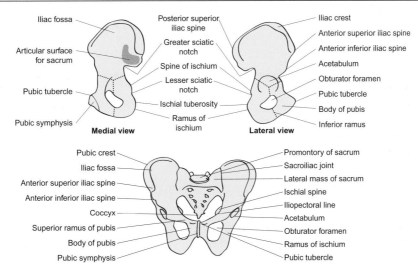

Answers

13. 1 – A, 2 – B, 3 – J, 4 – H, 5 – I, 6 – G, 7 – C, 8 – F, 9 – D, 10 – K, 11 – E
14. 1 – E, 2 – D, 3 – F, 4 – J, 5 – B, 6 – I, 7 – H, 8 – C, 9 – G, 10 – A
15. See explanation

16. The following muscles may be considered as either (1) part of the pelvic floor, (2) part
    of the muscular walls of the pelvis or (3) neither. Group them accordingly

Options

A. Coccygeus
C. Levator ani
E. Piriformis

B. Obturator internus
D. Rectus abdominis

17. Concerning the pelvis. Label the diagram below with the options provided

Options

A. Obturator internus muscle
C. Coccygeus muscle
E. Anterior inferior iliac spine
G. Lesser sciatic notch

B. Anterior superior iliac spine
D. Piriformis muscle
F. Obturator foramen
H. Levator ani muscle

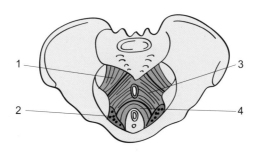

EXPLANATION: MUSCULAR WALLS AND FLOOR OF THE PELVIS

The walls of the pelvic cavity consist of the pubic symphysis anteriorly and sacrum posteriorly, with the **obturator internus** and **piriformis** muscles on the adjacent lateral sides. Originating from the bone surrounding and the membrane covering the obturator foramen, the obturator internus passes posteriorly until it reaches the lesser sciatic notch. Turning through 90 degrees, the muscle then travels out of the pelvis to insert upon the greater trochanter of the femur. The piriformis arises from the front of the three central segments of the sacrum. From here, it passes laterally out of the pelvis, inserting into the posterior part of the greater trochanter. Parietal pelvic fascia lines the walls of the pelvis and is named according to the muscle it overlies.

Supporting the pelvic viscera is the **pelvic diaphragm**. This is the floor of the pelvic cavity with the **levator ani** muscle at the front and the **coccygeus muscle** at the back. The fibres of the levator ani originate from the pubis anteriorly, the fascia of the obturator internus muscle from the side and the ischial spine posteriorly. From these attachments they converge across the midline with their corresponding muscle from the other side. However, the muscle is incomplete at the midline, with the **urogenital hiatus** (anteriorly) allowing passage of the urethra (and vagina in females) and the **anal canal** traversing posteriorly. The coccygeus muscle does not meet its counterpart from the opposite side as both muscles arise from their respective ischial spines and attach to the lateral surfaces of the sacrum and coccyx.

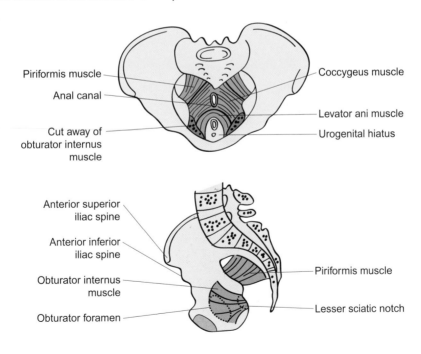

Piriformis muscle

Anal canal

Cut away of obturator internus muscle

Coccygeus muscle

Levator ani muscle

Urogenital hiatus

Anterior superior iliac spine

Anterior inferior iliac spine

Obturator internus muscle

Obturator foramen

Piriformis muscle

Lesser sciatic notch

Answers

16. 1 – B and E, 2 – A and C, 3 – D
17. 1 – D, 2 – A, 3 – C, 4 – H, 5 – B, 6 – E, 7 – F, 8 – G

18. Write short notes on the pelvic arteries

19. Using the options below label the nerves and blood vessels on the following diagram

A. Right common iliac artery
B. Pudendal nerve
C. Obturator nerve
D. Umbilical artery
E. Superior vesical artery
F. Right external iliac artery
G. Obturator artery
H. Right internal iliac artery
I. Lumbosacral trunk
J. Second sacral nerve
K. Left common iliac artery

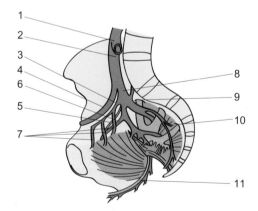

EXPLANATION: PELVIC ARTERIES AND NERVES

At the level of the fourth lumbar vertebra the abdominal aorta divides into the **common iliac arteries**, which in turn divide at the pelvic brim into **internal and external iliac arteries**. The internal iliacs perfuse the viscera of the pelvis, perineum and gluteal region, with the external iliacs supplying the remainder of the lower limbs. Branches of the internal iliac arteries within the pelvis show some variation between the sexes. However, in terms of the musculoskeletal system it is important to understand the anterior and posterior divisions of the artery that are common to both males and females. In this case, the anterior division gives off parietal (body wall and lower limb) branches; **obturator, internal pudendal and inferior gluteal arteries**. The posterior division gives off parietal branches only; **iliolumbar, lateral sacral and superior gluteal arteries (18)**.

The pelvis is innervated mainly by sacral and coccygeal nerves. Important nerves that innervate muscles of the pelvis are tabulated below.

Nerve	Origin	Motor distribution
Nerve to obturator internus	L5–S1, S2	Obturator
Nerve to piriformis	(S1), S2, S3	Piriformis muscle
Perineal	(S3), S4	Levator ani muscle
Coccygeal	S4, S5	Coccygeus muscle
Pudendal	(S1) S2, S3	Muscles of the perineum; skin of the penis and scrotum in males, and clitoris, labia majora and vagina in females

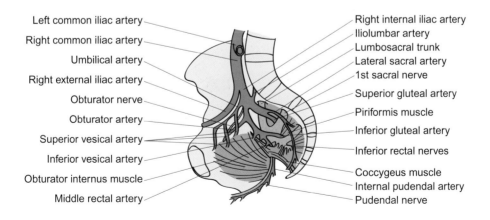

Left common iliac artery
Right common iliac artery
Umbilical artery
Right external iliac artery
Obturator nerve
Obturator artery
Superior vesical artery
Inferior vesical artery
Obturator internus muscle
Middle rectal artery

Right internal iliac artery
Iliolumbar artery
Lumbosacral trunk
Lateral sacral artery
1st sacral nerve
Superior gluteal artery
Piriformis muscle
Inferior gluteal artery
Inferior rectal nerves
Coccygeus muscle
Internal pudendal artery
Pudendal nerve

Answers
18. See explanation
19. 1 – K, 2 – A, 3 – D, 4 – F, 5 – C, 6 – G, 7 – E, 8 – H, 9 – I, 10 – J, 11 – B

20. Consider the perineum

    a. Define the bones that form the pelvic outlet

    b. Name two major structures that traverse the perineum

21. Match the options with the most accurate description from the list below

    A. Surrounded by an external and internal sphincter

    B. Posterior part of the perineum

    C. Made of three layers

    D. In males it contains the corpus spongiosum

    E. Forms one of the walls of the perineum

    1. Superficial perineal pouch

    2. Anal canal

    3. Urogenital triangle

    4. Obturator internus

    5. Anal triangle

22. Which of the following is NOT part of the superficial perineal pouch?

    A. Clitoris

    B. Inguinal ligament

    C. Labia majora

    D. Corpus spongiosum

    E. Bulbospongiosum

EXPLANATION: THE PERINEUM

The perineum lies below the pelvic diaphragm. Its walls are formed by the **obturator internus** muscles later-ally and the **bones** and **ligaments** that define the pelvic outlet, the bones being the **coccyx, symphysis pubis** and **ischium** on either side **(20a)**. The inferior boundary of the region is formed by the skin covering the area.

The perineum is subdivided into a posterior **anal triangle** and an anterior **urogenital triangle**. The anal tri-angle contains the **anal canal**. This is surrounded by the internal sphincter (smooth muscle – involuntary), which in turn is surrounded by the external sphincter (striated muscle – voluntary). The urogenital triangle has **three layers**. From superior to inferior, these are a superior fascia, a muscular layer (which includes the external sphincter of the urethra and deep transverse perineal muscles) and an inferior perineal membrane. The muscular layer is termed the **deep perineal pouch**, below which is the **superficial perineal pouch**. In males the superficial pouch contains the fibrous (corpus spongiosum and corpora cavernosa) and muscular (bul-bospongiosus and ischiocavernosus muscles) components of the penis. In females, equivalent structures that form the clitoris, labia (majora and minora) with associated fat, blood vessels and nerves are also part of the superficial perineal pouch **(20b)**.

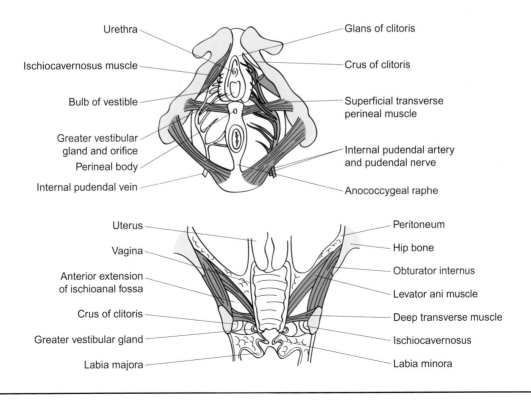

Urethra
Ischiocavernosus muscle
Bulb of vestible
Greater vestibular gland and orifice
Perineal body
Internal pudendal vein

Glans of clitoris
Crus of clitoris
Superficial transverse perineal muscle
Internal pudendal artery and pudendal nerve
Anococcygeal raphe

Uterus
Vagina
Anterior extension of ischioanal fossa
Crus of clitoris
Greater vestibular gland
Labia majora

Peritoneum
Hip bone
Obturator internus
Levator ani muscle
Deep transverse muscle
Ischiocavernosus
Labia minora

Answers
20. See explanation
21. 1 – D, 2 – A, 3 – C, 4 – E, 5 – B
22. B

23. Concerning abdominal hernias

a. Approximately 10 per cent of all hernias are inguinal
b. They are more likely to occur during pregnancy
c. They may be complicated by bowel obstruction
d. Femoral hernias are more common in men than women
e. Strangulated hernias must not be operated on

24. The diagram below depicts various hernias. Label it using the options provided

Options

A. Femoral
B. Inguinal (indirect)
C. Incisional
D. Umbilical
E. Inguinal (direct)

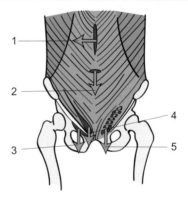

25. Match the following options with the cases described

Options

A. Rectocele   B. Urethrocele
C. Enterocele   D. Urethral prolapse
E. Cystocele   F. Rectal prolapse

1. Protrusion of the urethra into the vagina
2. Protrusion of the peritoneum or bowel into the vagina
3. Protrusion of the rectum through the anus
4. Protrusion of the urethra through the urethral opening
5. Protrusion of the rectum into the vagina
6. Protrusion of the bladder into the vagina

EXPLANATION: CLINICAL SCENARIOS

ABDOMINAL HERNIAS

An **abdominal hernia** is the **protrusion of peritoneum which may contain part or all of a viscus, from its normal position, through a weakened area in the abdominal wall**. Coughing, pregnancy and heavy lifting are all causative factors, as they may compress the abdominal cavity.

Inguinal hernias account for approximately 80 per cent of all hernias and are more common in men. On the other hand, **femoral hernias** are more common in women, and are more likely to occur during pregnancy. If left untreated, a hernia involving bowel may become **incarcerated** (cannot be pushed back into place) and obstruction of the bowel may occur. Also, such a hernia may become **strangulated**, where there is constriction of the blood supply to the bowel. A **strangulated hernia** is a **surgical emergency** as tissue necrosis and gangrene may ensue.

Type of hernia	Location
Incisional	At the site of a previous surgical procedure
Umbilical	Umbilical ring
Femoral	Upper medial thigh or just below inguinal ligament. Points towards leg
Indirect inguinal	Occurs at the internal (deep) ring and passes through the inguinal canal and then the external (superficial) ring. Points towards the groin
Direct inguinal	Pushes through posterior wall of inguinal triangle and then the external ring

PELVIC FLOOR PROLAPSE

A weakness in the supporting tissues and/or muscles of the pelvic floor may allow organs within the pelvis to herniate into the vagina. This is termed a genital prolapse, and is named after the tissue or organ that protrudes into the vagina: urethra – urethrocele; bladder – cystocele; uterus – uterine prolapse; peritoneum or small bowel – enterocele; and rectum – rectocele. Also structures in this region may herniate through passages other than the vagina; for example rectal prolapse through the anus and urethral prolapse through the urinary outlet.

The causes of prolapse are divided into factors that increase intra-abdominal pressure (for example childbirth and obesity) and factors that weaken or damage the pelvic floor (for example neuropathy and traumatic muscle tears). Treatment of a prolapse may involve the insertion of a pessary (to hold the prolapse) or reconstructive surgery. Exercising of the pelvic floor muscles has been shown to reduce the risk of prolapse.

Answers

23. F T T F F
24. 1 – C, 2 – D, 3 – A, 4 – E, 5 – B
25. 1 – B, 2 – C, 3 – F, 4 – D, 5 – A, 6 – E

EXPLANATION: MUSCLES OF RESPIRATION

The table below states the muscles involved in respiration:

	Quiet	Forced (in addition to muscles involved in quiet respiration)
Inspiration	Diaphragm (contraction) Intercostal muscles Scalene muscles	Sternocleidomastoid muscle Pectoralis muscles Serratus anterior muscle Erector spinae Quadratus lumborum
Expiration	Elastic recoil of lungs Intercostal muscles Diaphragm (relaxation)	Abdominal muscles Latissimus dorsi

SECTION 4

THE VERTEBRAL COLUMN

THE VERTEBRAL COLUMN

---

1. **True or false.** In the vertebral column

    a. There are seven cervical vertebrae
    b. There are five thoracic vertebrae
    c. There are twelve lumbar vertebrae
    d. The sacral vertebrae are fused
    e. The coccyx lies at the caudal end of the vertebral column

---

2. **Label regions 1–5 on the diagram below**

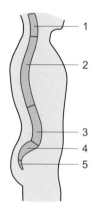

---

3. **In the neonate**

    a. The curvature of the vertebral column is the same as that of adults
    b. The vertebral column is anteriorly flexed like the letter 'C'
    c. The secondary cervical curvature develops as children learn to walk
    d. The secondary lumbar curvature develops as children learn to hold their head up
    e. Both secondary curvatures are concave posteriorly

EXPLANATION: INTRODUCTION TO THE VERTEBRAL COLUMN

The **vertebral column** serves four functions: it **supports the skull**, provides **anchorage for the ribs**, provides **anchorage for shoulders and hips** and **protects the spinal cord**. It is divided into five main regions as shown in the figure: **cervical (seven vertebrae)**, **thoracic (twelve vertebrae)**, **lumbar (five vertebrae)**, **sacral (five fused vertebrae)** and **coccyx (2)**. The coccyx lies at the end of the vertebral column and comprises up to four small fused remnants of vertebrae. Note the **kyphosis** of the thoracic part and **lordosis** of the lumbar part of the adult vertebral column are shown in lateral view in the figure.

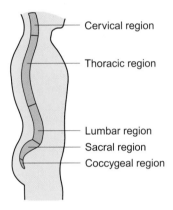

Cervical region

Thoracic region

Lumbar region
Sacral region
Coccygeal region

In **neonates and infants** the vertebral column is **anteriorly flexed** like a letter 'C'. This is known as the **primary curvature**. Children develop two **secondary curvatures**: the **cervical** curvature as they learn to hold their head up and the **lumbar** curvature as they learn to walk. Both secondary curvatures are concave posteriorly. A lateral view of the neonate vertebral column is shown below; note the anterior flexion resembling the letter 'C'.

Answers

1. T F F T T
2. See figure
3. F T F F T

4. Label structures 1–5 on the superior view of the cervical vertebra below

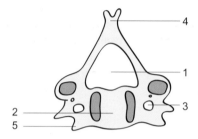

5. Regarding the first and second cervical vertebrae

   a. They are known respectively as the atlas and axis
   b. The axis articulates directly with the skull
   c. The atlas has no body
   d. The odontoid process of the axis assists rotation of the head
   e. The superior articular facets of the atlas assist rotation of the skull

6. True or false. With regard to cervical vertebrae

   a. The spinous process of C7 is the longest and is known as vertebra prominens
   b. The spinous processes become more bifid the lower they are down the vertebral column
   c. The vertebral arteries run through the foramina transversaria of C1–C6
   d. The anterior tubercles of C5 are known as the carotid tubercles
   e. There is no disc between the axis and C3

EXPLANATION: THE CERVICAL VERTEBRAE

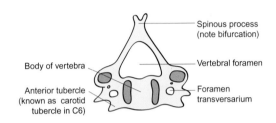

- Spinous process (note bifurcation)
- Vertebral foramen
- Foramen transversarium
- Body of vertebra
- Anterior tubercle (known as carotid tubercle in C6)

The **first two cervical** vertebrae are known respectively as the **atlas and** the **axis**. The **atlas** has **no body** and **articulates directly with the skull**; it has no disc. The **axis** does have a body from which a vertical peg of bone, the **dens** or **odontoid process**, projects. In the figure opposite, note the dens projecting upwards into the anterior aspect of the vertebral foramen of the atlas. It is held in place by the transverse ligament.

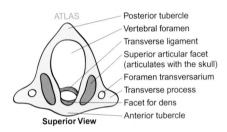

ATLAS
- Posterior tubercle
- Vertebral foramen
- Transverse ligament
- Superior articular facet (articulates with the skull)
- Foramen transversarium
- Transverse process
- Facet for dens
- Anterior tubercle

Superior View

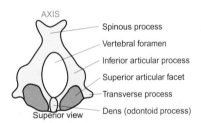

AXIS
- Spinous process
- Vertebral foramen
- Inferior articular process
- Superior articular facet
- Transverse process
- Dens (odontoid process)

Superior view

The **atlas supports** the **skull**. Its superior articular facets allow **flexion and extension** of the skull. The **axis** articulates with the atlas via the superior articular facets on the former and the inferior articular facets on the latter as well as the dens. This **allows the atlas and skull to rotate** around the axis (5d).

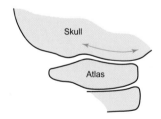

Skull

Atlas

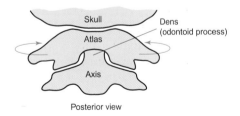

Skull

Atlas

Axis

Dens (odontoid process)

Posterior view

C1–C7 are all joined together by discs. They all have **foramina transversaria**. The **vertebral arteries** enter the foramina of C6 and ascend towards the skull. The **spinous processes** of these vertebrae are **bifid**, but become less so the lower they are down the vertebral column. The **spinous process of C7 – vertebra prominens – is** the **largest** and therefore most prominent. The anterior and posterior tubercles of the cervical vertebrae are where some of the muscles of the neck insert. The anterior tubercles of C6 are known as the carotid tubercles. **Flexion, extension and rotation** all occur at the articular facets of the **zygopophyseal joints** (the joints formed by intervertebral articular facets).

Answers

4. See figure
5. T F T T F
6. T F T F F

7. Label structures 1–5 on the views of the thoracic vertebra below

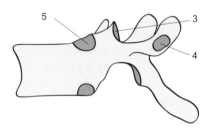
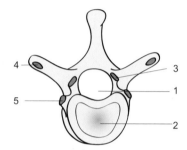

8. With regard to thoracic vertebrae

    a. Each vertebra articulates with at least one pair of ribs
    b. The anteroposterior diameter is larger than the transverse
    c. The vertebral foramina are circular
    d. No flexion or extension is allowed at the zygapophyseal joints
    e. Rotation is allowed at the zygapophyseal joints

9. Label structures 1–5 on the superior view of a lumbar vertebra below

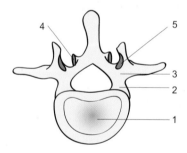

10. True or false. With regard to lumbar vertebrae

    a. Lumbar vertebrae are larger than cervical but smaller than thoracic
    b. The spinous processes gradually disappear from L1 to L5
    c. No flexion or extension are permitted at the zygapophyseal joints
    d. Minimal rotation is permitted at the zygapophyseal joints
    e. The fifth lumbar vertebra does not articulate by means of an intervertebral disc with the sacrum

EXPLANATION: THE THORACIC AND LUMBAR VERTEBRAE

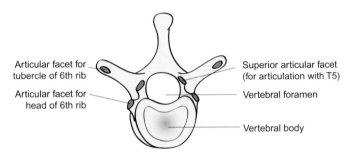

Articular facet for
tubercle of 6th rib

Articular facet for
head of 6th rib

Superior articular facet
(for articulation with T5)

Vertebral foramen

Vertebral body

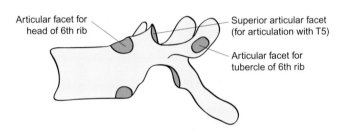

Articular facet for
head of 6th rib

Superior articular facet
(for articulation with T5)

Articular facet for
tubercle of 6th rib

Thoracic vertebrae are all joined together by discs. **T1** articulates **with rib 1 and T11 and T12 with ribs 11 and 12**. However, **all the other ribs** articulate **with their respective vertebrae and the one above** (the head of the rib therefore straddles the intervertebral disc).

In thoracic vertebrae, the **anteroposterior diameter is roughly the same as the transverse**. This results in **circular vertebral foramina** in that region of the vertebral column.

It is worth noting that T1–T4 have some cervical features while T9–T12 have some lumbar features. It is only the middle four vertebrae (T5–T8) that are truly and typically thoracic.

Although **rotation** is the **main movement allowed** at the zygapophyseal joints of **thoracic vertebrae**, flexion and extension also occur.

Lumbar vertebrae are all joined together by discs. L5 articulates with the sacrum via a disc as well. **Lumbar vertebrae** are **larger** than cervical and thoracic vertebrae. Their **transverse diameter is greater than the anteroposterior**. This gives them an **oval shape**.

The zygapophyseal joints between lumbar vertebrae **mainly** allow **flexion** and **extension**. Rotation occurs to a lesser extent.

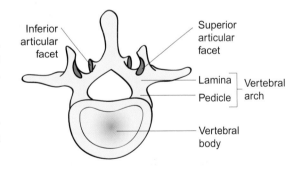

Inferior
articular
facet

Superior
articular
facet

Lamina — Vertebral
arch
Pedicle

Vertebral
body

11. Briefly discuss the main differences between cervical, thoracic and lumbar vertebrae

12. Consider the sacrum

   a. It consists of four fused vertebrae
   b. It is triangular in shape
   c. It articulates with the hip bones of the pelvis through the sacroiliac joints
   d. The body weight is transmitted via the sacroiliac joints
   e. It is the same size in females as it is in males

13. How many bones form the coccyx, what shape is it and how does it articulate with the sacrum?

EXPLANATION: THE VERTEBRAE, SACRUM AND COCCYX

Cervical vertebrae have **foramina transversaria** which are absent in thoracic and lumbar vertebrae. **Thoracic** vertebrae have **facets** for articulation with the **heads of the ribs**, a feature absent in cervical and lumbar vertebrae. **Lumbar** vertebrae are **larger** on the whole than thoracic and cervical vertebrae. The **spinous processes of cervical** vertebrae are **bifid. All three types of vertebrae** allow various amounts of **flexion, extension and rotation** at the articular facets of their **zygapophyseal joints (11)**.

The **sacrum is triangular** in shape. It is the result of the **fusion of five sacral bodies** – essentially vertebrae **(S1–S5)**. This fusion occurs throughout puberty and may still be taking place well into the twenty-fifth year of life.

S1–S3 all have spines. **In S4** and **S5** these **spines** are **absent**. In fact the bodies of these two vertebrae are deficient posteriorly – the **sacral hiatus** as shown in the posterior view of the sacrum below.

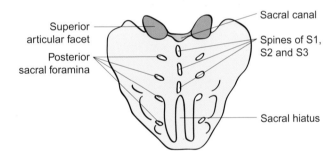

Superior articular facet —
Posterior sacral foramina —
Sacral canal
Spines of S1, S2 and S3
Sacral hiatus

Body weight is transmitted **via sacroiliac joints**.

The **sacrum is wider** and the body narrower **in females** than in males.

The coccyx consists of up to **four fused bones** (remnants of vertebrae). It is **triangular** in shape. It **articulates** with the **sacrum via** an **intervertebral disc and** laterally via **two small synovial joints (13)**.

Answers

11. See explanation
12. F T T T F
13. See explanation

14. True or false. The intervertebral discs

a. Are fibrocartilaginous structures
b. Allow no movement of the vertebrae they connect
c. Allow free movement of the vertebrae they connect
d. Never prolapse
e. Act as shock absorbers in the vertebral column protecting the brain and spinal cord from sudden impacts

15. Consider the intervertebral discs

a. They consist of two parts: the annulus fibrosus and the nucleus pulposus
b. The nucleus pulposus surrounds the annulus fibrosus
c. The annulus fibrosus consists of fibrous tissue
d. The nucleus pulposus is a gelatinous ball
e. The nucleus pulposus becomes less gelatinous with age

EXPLANATION: THE INTERVERTEBRAL DISCS

The intervertebral discs are **fibrocartilaginous** structures that allow **limited movement** of the vertebrae they connect. They may therefore each be regarded as **part of a symphyseal joint**. As well as allowing a limited range of movement of the vertebral column, they also act as **shock absorbers** and assist in protecting the brain and spine from sudden impact and injury.

Each intervertebral disc consists of **two parts**: the **annulus fibrosus** and the **nucleus pulposus**. The **former** consists of **fibrous tissue** and exists at the **periphery** of the disc, whereas the **latter** is a **gelatinous** ball at the **centre** of the disc.

It is worth noting that the nucleus pulposus becomes less gelatinous with advancing age. It can herniate through splits in the annulus and may, as a result, compress spinal nerves or even the spinal cord itself. This is known as a **disc prolapse**. A superior view of a transverse section of an intervertebral disc is shown below.

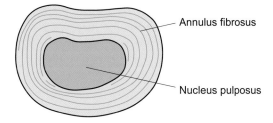

Annulus fibrosus

Nucleus pulposus

---

14. T F F F T
15. T F T T T

16. With regard to the ligaments of the vertebral column

    a. The ligamentum flavum connects the laminae of adjacent vertebrae
    b. The interspinous ligaments connect the inferior and superior borders of adjacent transverse processes
    c. The supraspinous ligaments extend inferiorly as far as L4 or L5
    d. The interspinous and supraspinous ligaments are continuous
    e. The anterior and posterior longitudinal ligaments unite adjacent transverse processes

17. Match the ligaments to the correct statements from the options below

Options

    A. Extend between the sacrum and coccyx
    B. Extend from the ilia to the transverse processes of L5
    C. Extend from the transverse processes of L3 to the ischial spines
    D. Extend along all the transverse processes of the lumbar vertebrae
    E. Extend from the sacrum and coccyx to the ischial spines
    F. Extend between the ischial spines and the ischial tuberosities
    G. Extend from the ilia to the transverse processes of L4
    H. Extend along the anterior surface of the coccyx
    I. Extend from the sacrum, ilium and coccyx to the tuberosities of the ischia
    J. Extend posteriorly from the sacrum to the iliac tuberosities and posterior superior iliac spines

    1. The iliolumbar ligaments
    2. The posterior sacroiliac ligaments
    3. The sacrotuberous ligaments
    4. The sacrospinous ligaments
    5. The sacrococcygeal ligaments

EXPLANATION: THE LIGAMENTS OF THE VERTEBRAL COLUMN

The **ligamentum flavum** and the **interspinous** and **supraspinous** ligaments are all continuous. The **ligamentum flavum** unites superior and inferior borders of adjacent laminae, while the **interspinous** ligaments unite the borders of adjacent spinous processes. The **supraspinous** ligaments run over the tips of the spinous processes and extend as far inferiorly as L4 or L5. In the cervical part of the vertebral column they are known collectively as the **nuchal ligament** and attach superiorly to the nuchal line on the occipital bone of the skull. The **anterior and posterior longitudinal ligaments** unite adjacent vertebral bodies anteriorly and posteriorly. A sagittal section showing the vertebrae and ligaments is shown below.

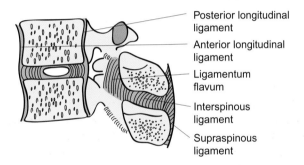

The **iliolumbar** ligaments extend from the ilia to the transverse processes of **L5**. A lumbosacral portion of the iliolumbar ligament extends inferiorly and blends with the anterior sacroiliac ligaments. The **posterior sacroiliac** ligaments extend posteriorly from the sacrum to the iliac tuberosities and posterior superior iliac spines. The anterior sacroiliac ligaments connect the anterior aspect of the joint. The **sacrotuberous and sacrospinous** ligaments unite the sacrum and coccyx with respectively the ischial tuberosities and spines. The sacrococcygeal ligaments are the equivalent of the anterior and posterior longitudinal ligaments.

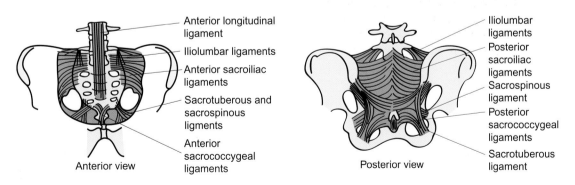

Anterior view

Posterior view

Answers

16. T F T T F
17. 1 – B, 2 – J, 3 – I, 4 – E, 5 – A

18. Consider the musculature of the back

    a. There are two layers of muscles: extrinsic and intrinsic
    b. The prevertebral muscles are extrinsic
    c. The psoas major is a prevertebral muscle
    d. The postvertebral muscles are collectively known as erector spinae
    e. The intrinsic muscles are the only ones in the body to be supplied by posterior rami of the spinal nerves

19. Regarding the musculature of the back

    a. The trapezius inserts superiorly into the nuchal ligament
    b. The trapezius, latissimus dorsi, levator scapulae and rhomboids all help attach the upper limb to the trunk
    c. The serratus posterior extends from the vertebral spines to the ribs and is a muscle of respiration
    d. The serratus posterior is divided into two parts: medial and lateral
    e. The levator scapulae, rhomboids and serratus posterior superior are all deep to the trapezius

EXPLANATION: THE MUSCLES OF THE VERTEBRAL COLUMN AND BACK

Muscles of the back may be subdivided into **extrinsic** and **intrinsic**. In turn, the **intrinsic muscles** may themselves be subdivided into **prevertebral or postvertebral** muscles. The **psoas major** is an example of a **prevertebral muscle**. There are two major groups of **postvertebral muscles**: the **erector spinae** muscles and deep to them the **transverso-spinalis** muscles.

Similarly, the **extrinsic muscles** may be subdivided into **superficial and deep** muscles. The **superficial** muscles are the **trapezius and latissimus dorsi**, and the **deep** ones are the **levator scapulae, rhomboids, and serratus posterior superior and inferior**. These muscles are shown in the posterior view of the back below, but note that the serratus posterior superior is not shown. It runs anterior to the rhomboids, inferolaterally from the spines of the cervical and thoracic vertebrae to the ribs.

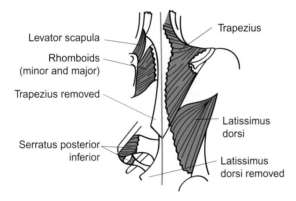

The **trapezius** inserts into the nuchal ligament, the spinous processes and the spine of the scapula. All of the extrinsic muscles except the serratus posterior help attach the upper limb to the trunk. The serratus **posterior** is a **muscle of respiration**. It extends **from** the **vertebral spines to** the **ribs**.

The **intrinsic muscles** are the **only** muscles in the body to be supplied by the **posterior rami** of spinal nerves, while the **extrinsic** muscles are **supplied by anterior rami**.

Answers
18. T F T F T
19. T T T F T

20. Consider the vasculature of the vertebral column and spinal cord

    a. The spinal cord receives its blood supply via three longitudinal spinal arteries
    b. The three spinal arteries are all anterior to the spinal cord
    c. The spinal arteries are all branches of the vertebral arteries
    d. Spinal veins also exist in the same distribution as that of the spinal arteries
    e. The spinal arteries also supply the vertebrae with blood

21. On the diagram of the spinal cord and meninges below label structures 1–5

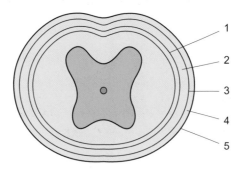

EXPLANATION: THE BLOOD SUPPLY TO THE VERTEBRAL COLUMN AND SPINAL CORD

The spinal cord derives its blood supply from **three longitudinal spinal arteries** – one anterior which runs along the ventral surface of the cord in the midline, and two posterior which run on the posterolateral surfaces of the cord.

The two **posterior spinal arteries** arise from the **vertebral arteries** or **posterior inferior cerebellar arteries**, whereas the **anterior spinal artery** is derived in a Y-shaped configuration from the **basilar artery**.

The anterior and posterior spinal arteries are reinforced at various intervals by anterior and posterior segmental medullary arteries. Where these arteries do not occur, radicular arteries supply the dorsal and ventral roots.

Vertebrae also derive their blood supply from **spinal arteries**.

The **distribution of the veins is similar** to that of the arteries.

The vertebrae possess a venous plexus known as the **valveless vertebral venous plexus of Batson**. It consists of an **internal or epidural** venous plexus **and an external plexus. Blood drains from** the **internal** plexus **to** the **external** one **and into the lumbar, azygos** and **hemiazygos veins**.

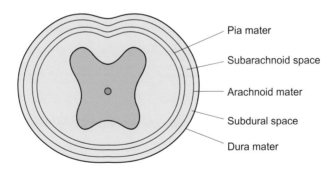

- Pia mater
- Subarachnoid space
- Arachnoid mater
- Subdural space
- Dura mater

22. Concerning the spinal cord and nerves

A. At what vertebral level does the spinal cord end?
B. What is the collection of spinal nerves that carry on down the spinal canal known as?
C. What structures do they supply?

23. True or false. Regarding spinal nerves

a. There are 31 pairs of spinal nerves
b. Each spinal nerve contains both anterior and posterior roots
c. The posterior roots contain efferent neurons whereas the anterior roots contain afferent neurons
d. C1–C7 spinal nerves exit from the vertebral canal above their corresponding vertebrae
e. T1–S5 all exit above their corresponding vertebrae

EXPLANATION: THE SPINAL CORD AND SPINAL NERVES

The **spinal cord ends at** level **L2** where it is known as the conus medullaris **(22A)**. The **cauda equina** is the collection of lumbar, sacral and coccygeal nerves that carry on down the spinal canal **from** level **L2 (22B)**. These spinal nerves supply the **lower limbs** and **perineum (22C)**. The tapering terminal section of the cauda equina is known as the **filum terminale**.

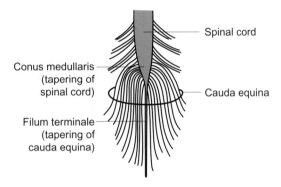

There are **31 pairs of spinal nerves** (eight cervical, twelve thoracic, five lumbar, five sacral and one coccygeal).

Each spinal nerve has **posterior** and **anterior roots**. The **posterior** roots are **afferent** – sensory fibres from the periphery to the spinal cord and brainstem. The **anterior** roots are **efferent** – motor fibres from spinal cord and brainstem to the periphery.

Spinal nerves **C1–C7 exit** from the vertebral canal **above** their **corresponding vertebrae**, whilst **C8 exits below** the seventh vertebra. **T1–S5** all **exit below** their corresponding vertebrae.

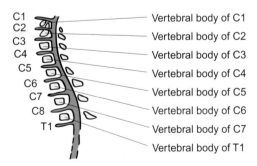

Answers

22. See explanation
23. T T F T F

24. Match the spinal nerves below with the muscles (myotomes) they supply

Options

1. Deltoid
2. Triceps
3. Quadriceps
4. Biceps
5. Gastrocnemius

A. C5
B. C7 and C8
C. C6
D. C5 and C6
E. L1, L2 and L3
F. L4, L5, S1 and S2
G. C6, C7 and C8
H. S1 and S2
I. L3 and L4
J. L5 and S1

25. With regard to myotomes of the shoulder and back

a. The trapezius is supplied by C2
b. The rhomboids are supplied by C3
c. C5 supplies the supraspinatus
d. The infraspinatus is supplied by C5 and C6
e. The latissimus dorsi is supplied by C7 only

EXPLANATION: MYOTOMES

Each spinal nerve has a posterior and an anterior root. The **posterior root** is **sensory** and **supplies a dermatome**, while the **anterior** root **is motor** and **supplies a myotome**.

Muscles usually consist of **more than one myotome** and are, as a result, supplied by more than one spinal nerve.

The task of remembering the exact spinal innervation of all muscles is a colossal one. Below are the most important:

Muscle	Innervation
Upper limb	
Deltoid	C5 and C6
Biceps	C5 and C6
Brachialis	C5 and C6
Triceps	C6, C7 and C8
Brachioradialis	C5 and C6
Pronator teres	C6 and C7
Supinator	C5, C6 and C7
Lower limb	
Iliopsoas	L1–L4
Quadriceps	L3 and L4
Biceps femoris	L4, L5, S1 and S2
Tibialis anterior	L4 and L5
Soleus	L4, L5, S1 and S2
Gastrocnemius	L4, L5, S1 and S2
Back	
Supraspinatus	C5
Infraspinatus	C5 and C6
Subscapularis	C5 and C6
Trapezius	C3 and C4
Latissimus dorsi	C6, C7 and C8
Rhomboids	C4 and C5

Answers

24. 1 – D, 2 – G, 3 – I, 4 – D, 5 – F
25. F F T T F

26. Match the dermatomes 1–5 on the diagram below with their respective spinal nerves from the list

Options

A. C5
B. S3
C. L4
D. T5
E. L1
F. L5
G. T10
H. L3
I. C3
J. S5

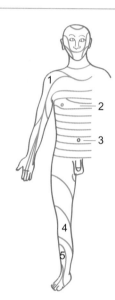

27. With regard to dermatomes

a. Dermatomes are segments of skin supplied by the sensory fibres of a single spinal nerve
b. There is considerable overlap of adjacent dermatomes
c. Dermatomes are segments of skin supplied by the motor fibres of a single anterior root
d. The head consists of a single dermatome
e. Dermatomes are supplied by a single spinal nerve

EXPLANATION: DERMATOMES

Dermatomes are segments of skin supplied by a **single posterior root** – therefore **the sensory fibres of a single spinal nerve**. There is considerable overlap of adjacent dermatomes and some individual variation. The approximate distribution of dermatomes is shown in the figures below.

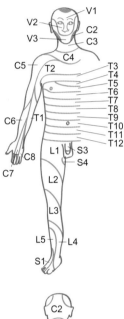

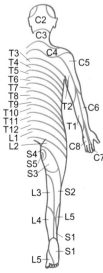

26. 1 – A, 2 – D, 3 – G, 4 – H, 5 – F
27. T T F F F

28. Case study

A 40-year-old man presents to his GP with shooting pain and loss of sensation down the left leg. On further questioning it becomes evident that the pain initially started in the back about four days ago. The onset was sudden and was preceded by strenuous exercise at the gym.

What is the most likely diagnosis (only one option is correct)?

Options

    A. Disc prolapse
    B. Acute spinal cord compression
    C. Osteoarthritis
    D. Rheumatoid arthritis
    E. Osteoporosis

29. Case study

A 35-year-old motorist was involved in a road traffic accident whereby her car was hit from the back. Soon after the accident she complained of pain and stiffness in her neck. This is still ongoing six months later.

What is the most likely diagnosis (only one option is correct)?

Options

    A. Fracture of the odontoid process
    B. Rupture of the nuchal ligament
    C. Fracture of C7
    D. A cervical strain ('whiplash') injury
    E. Acute spinal cord compression

EXPLANATION: CLINICAL SCENARIOS

28. Acute disc prolapse is a condition whereby the nucleus pulposus herniates through the annulus fibrosus. It most commonly occurs in young people, 20–40 years old, and is a common cause for back pain in that age group. The pain is often in the lower back. It is severe, of sudden onset (usually following strenuous activity such as lifting heavy loads) and often radiates to the buttock and leg if a spinal root is impinged. There may be associated sensory loss in the distribution of the impinged spinal root.

On examination, the patient may have a sideways tilt, known as a sciatic scoliosis, usually towards the painful side. **Forward flexion and extension** are usually **limited**. One of the **nerve stretch tests** is usually **positive** (sciatic or femoral).

The initial management is conservative (bed rest, pain killers and physiotherapy). If the pain does not resolve and/or symptoms get worse, the majority of patients may benefit from bed rest and traction for two weeks. In those that do not, an epidural injection of corticosteroids and local anaesthetic may help. If all else fails, chemonucleolisis (intradiscal injection of chymopapain which dissolves the nucleus) or discectomy (removal of part of or the whole of a disc) are very effective.

Recovery is usually complete after three months. Advice on heavy lifting can be preventative of future episodes.

29. Cervical strain or 'whiplash' injury are terms applied to a soft tissue injury, which usually follows sudden hyperextension of the neck. Although the exact pathology is still under debate, it is most likely that the injury consists of damage to the anterior longitudinal ligament and one or more cervical discs. The main complaints are those of pain and stiffness in the neck which may last years. Patients are offered analgesia and physiotherapy, however, no form of treatment has been proven to be of great value.

Answers

28. A
29. D

SECTION 5

THE UPPER LIMB

THE UPPER LIMB

1. Concerning the clavicle and the scapula

    a. The scapula covers the first to fifth ribs posteriorly
    b. The coracoid process articulates with the humeral head
    c. The clavicle articulates with the acromion process laterally
    d. The clavicle articulates with the manubrium medially
    e. The spine of scapula is continuous with the acromion process

2. Label the following diagrams of the pectoral girdle with the options provided

Options

A. Supraspinous fossa	**B.** Infraspinous fossa
C. Subscapular fossa	**D.** Glenoid fossa
E. Spine of scapula	**F.** Neck of scapula
G. Acromion process	**H.** Coracoid process
I. Humerus	**J.** Clavicle
K. Superior angle	**L.** Inferior angle

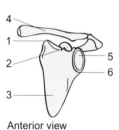

Anterior view

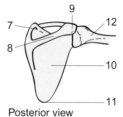

Posterior view

EXPLANATION: BONES OF THE UPPER LIMB (i)

The medial two-thirds of the **clavicle** are circular in section and convex anteriorly, while the lateral one-third is flattened and curved convex posteriorly. Medially the clavicle articulates with the **manubrium** at the **sternoclavicular joint**. Laterally it articulates with the **acromion process** of the scapula at the **acromioclavicular joint**. The **scapula** is a flattened, triangular bone. It covers the **second to seventh ribs** posteriorly. Laterally the **glenoid fossa** articulates with the humeral head.

The **coracoid process** projects anteriorly from the superior border of the scapula. The anterior surface of the scapula is called the **subscapular fossa**. The **spine of the scapula** divides the posterior surface into the **supraspinous fossa** superiorly and **infraspinous fossa** inferiorly. The spine of the scapula projects laterally as the **acromion process**.

Answers
1. F F T T T
2. 1 – K, 2 – H, 3 – C, 4 – J, 5 – D, 6 – F, 7 – A, 8 – E, 9 – G, 10 – B, 11 – L, 12 – I

3. **Label the following diagrams with the options provided. You may use the options once, more than once or not at all**

Options

A. Head of radius   B. Head of ulna   C. Head of humerus
D. Anatomical neck   E. Surgical neck   F. Capitulum
G. Shaft of humerus   H. Shaft of ulna   I. Shaft of radius
J. Lateral epicondyle   K. Medial epicondyle   L. Radial tuberosity
M. Coronoid process   N. Radial styloid process O. Trochlea notch
P. Greater tuberosity   Q. Lesser tuberosity   R. Olecranon process
S. Deltoid tuberosity   T. Trochlea

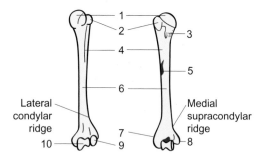

Lateral condylar ridge
Medial supracondylar ridge

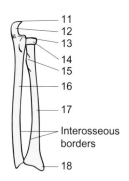

Interosseous borders

4. **Label the following diagram with the options provided. You may use the options once, more than once or not at all**

Options

A. Radius   B. Trapezoid
C. Triquetral   D. Trapezium
E. Lunate   F. Capitate
G. Hamate   H. Scaphoid
I. Pisiform   J. Distal phalanx
K. Proximal phalanx   L. Middle phalanx
M. Metacarpals

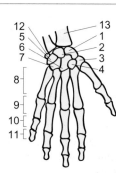

EXPLANATION: BONES OF THE UPPER LIMB (ii)

The rounded humeral head articulates with the **glenoid fossa** of the scapula. The **anatomical neck** separates the humeral head from the **greater** and **lesser tuberosity**. Between the tuberosities is the **intertubercular sulcus**. Distal to the anatomical neck is the surgical neck, where the humerus narrows to become the **shaft**. The **spiral groove** extends inferolaterally on the posterior aspect of the body (where the **radial nerve** courses). On the anterolateral aspect of the shaft there is a roughness known as the **deltoid tuberosity**. The distal end of the humerus consists of the **trochlea** medially and the round **capitulum** laterally. They articulate with the **olecranon** and **coronoid processes** of the ulna and the head of the radius, respectively. On either side are the **medial** and lateral **epicondyles**.

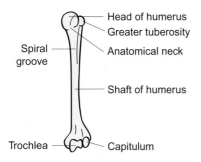

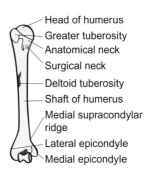

The **radial head** is at its proximal end, while the **ulnar head** is at its distal end. Distal to the neck of the radius is the **radial tuberosity**. Only the distal end of the **radius** articulates with the **scaphoid** and **lunate** carpal bones of the wrist. The lateral distal end of the radius and the medial distal end of the ulna form the prominent **styloid processes**.

Proximal row of **carpal bones**, from lateral to medial, consists of the **scaphoid**, **lunate**, **triquetrum** and the **pisiform**. The articulation between these bones and the radius is known as the **wrist joint**. It is a **synovial joint**. It is a **synovial joint**. Distal row of carpal bones, from lateral to medial, consists of the **trapezium**, **trapezoid**, **capitate**, and **hamate** (see figure on question 4). The carpal bones articulate with five metacarpal bones (**carpometacarpal joints**), forming the framework of the palm. Articulations between the metacarpal and the proximal phalanges (the **metacarpophalangeal joints**) form the knuckles. Each **finger** consists of a **proximal**, a **middle** and a **distal phalanx** (joined up by **interphalangeal joints**), while the **thumb** contains only a **proximal** and a **distal phalanx**.

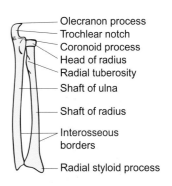

Answers
3. 1 – C, 2 – P, 3 – E, 4 – D, 5 – S, 6 – G, 7 – J, 8 – K, 9 – F, 10 – T, 11 – R, 12 – O, 13 – M, 14 – A, 15 – L, 16 – H, 17 – I, 18 – N
4. 1 – H, 2 – E, 3 – D, 4 – B, 5 – C, 6 – F, 7 – G, 8 – M, 9 – K, 10 – L, 11 – J, 12 – I, 13 – A

5. Concerning arteries of the upper limb

    a. The axillary artery continues from the subclavian artery and runs medial to the axillary vein

    b. The brachial artery bifurcates into the radial and ulnar arteries under the tricipital aponeuosis

    c. The radial artery can be felt at the anatomical snuffbox

    d. The ulnar artery passes through the carpal tunnel

    e. The ulnar artery runs lateral to the ulnar nerve in the distal two-thirds of the forearm

6. The following anteroposterior view of the right upper limb illustrates the anatomy of the venous system. Label it with the options provided. You may use the options once, more than once or not at all.

A. Medial cutaneous nerve of forearm profunda

B. Cephalic vein

C. Profunda brachii artery

D. Median nerve

E. Subclavian vein

F. Ulnar nerve

G. Median cubital vein

H. Basilic vein

I. Axillary vein

J. Brachial artery

K. Musculocutaneous nerve

L. Radial nerve

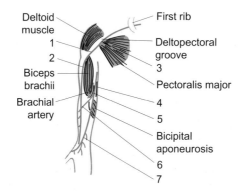

Deltoid muscle — First rib — Deltopectoral groove — 1 — 2 — 3 — Biceps brachii — Pectoralis major — Brachial artery — 4 — 5 — Bicipital aponeurosis — 6 — 7

EXPLANATION: VESSELS OF THE UPPER LIMB

The arterial system of the upper limb consists of the:

- **Axillary artery**: continues from the **subclavian artery** and commences at the lateral border of the first rib. The **brachial plexus** lies around the axillary artery above the level of **pectoralis minor**. The **axillary vein** lies medial to the axillary artery throughout its course. Branches of the axillary artery supply the shoulder and the chest wall
- **Brachial artery**: the axillary artery ends at the lower border of the **teres major** to become the brachial artery, which lies immediately under the deep fascia. It is crossed anteriorly by the **median nerve** in the mid-arm (lateral to medial). Its **profunda brachii** branch accompanies the radial nerve and takes part in the anastomosis around the elbow joint. The brachial artery divides into the **radial** and **ulnar arteries** under the **bicipital aponeurosis**
- **Radial artery**: is overlapped by the **brachioradialis**, and lies between it and the **flexor carpi radialis** distally. It then passes through the anatomical snuffbox where it can be palpated
- **Ulnar artery**: passes between the ulnar and radial heads of the **flexor digitorum superficialis** with the **median nerve**. Distally it runs lateral to the **ulnar nerve**. The ulnar artery passes superficial to the **flexor retinaculum** and not the carpal tunnel.

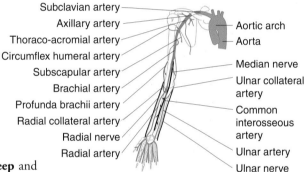

Subclavian artery — Axillary artery — Thoraco-acromial artery — Circumflex humeral artery — Subscapular artery — Brachial artery — Profunda brachii artery — Radial collateral artery — Radial nerve — Radial artery — Aortic arch — Aorta — Median nerve — Ulnar collateral artery — Common interosseous artery — Ulnar artery — Ulnar nerve

The **radial** and **ulnar arteries** are connected at the **deep** and **superficial palmar arches** distally. The deep palmar arch gives off three palmar metacarpal arteries and supplies the digits with the common palmar digital arteries from the superficial palmar arch. The radial and ulnar arteries also anastomose via the **anterior** and **posterior interosseous arteries** between the ulna and the radius, where the former supplies the flexor muscles and the latter supplies the extensor muscles. They are branches of the **common interosseous artery** (the first ulnar branch).

The venous drainage of the upper limb (see figure on question 6) consists of the:

- **Brachial veins**: these two veins begin at the elbow by union of the deep venae commitantes of the ulnar and radial arteries and end in the axillary vein
- **Cephalic vein**: derives from the lateral side of the dorsal venous network overlying the **anatomical snuffbox** (see page 103). The vein ascends the lateral then anterolateral aspects of the forearm and the superficial surface of the biceps of the arm, coursing through the **deltopectoral groove**. It pierces the clavipectoral fascia and drains into the **axillary vein**
- **Basilic vein**: derives from the medial aspect of the dorsal network, and ascends the medial then anteromedial aspects of the forearm and arm. It continues in the mid-arm as the **axillary vein** at the inferior border of the **teres major**
- **Medial cubital vein**: interconnects the cephalic and basilic veins in the cubital fossa anterior to the **bicipital aponeurosis**, where it is an important site for phlebotomy and peripheral venous access.

Answers

5. T T T F T
6. 1 – I, 2 – B, 3 – E, 4 – A, 5 – D, 6 – G, 7 – H

---

7. With regard to the brachial plexus, label the diagram with the options provided

Options

A. Phrenic nerve
B. Suprascapular nerve
C. Ulnar nerve
D. Median nerve
E. Radial nerve
F. Musculocutaneous nerve
G. Thoracodorsal nerve
H. Upper trunk
I. Middle trunk
J. Lower trunk
K. Long thoracic nerve
L. Vagus nerve

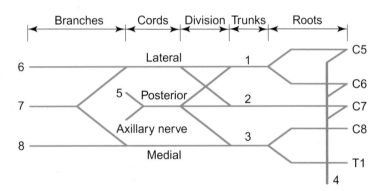

---

8. Concerning nerves of the upper limb

a. The axillary nerve winds round the head of the humerus
b. The axillary nerve provides motor supply to the deltoid muscle
c. The radial nerve supplies muscles in the flexor compartments of the upper limb
d. The musculocutaneous nerve passes through the two heads of the coracobrachialis
e. The musculocutaneous nerve descends in the arm between the brachialis and the biceps

---

9. Concerning nerves of the upper limb

a. The radial nerve gives branches to the triceps and brachioradialis
b. The median nerve gives rise to the posterior interosseous nerve
c. The median nerve supplies skin overlying the thenar eminence
d. The ulnar nerve runs on the lateral side of the brachial artery
e. The ulnar nerve pierces through the coracobrachialis

EXPLANATION: NERVOUS SYSTEM OF THE UPPER LIMB

The **brachial plexus** (C5 to T1) is a network of nerves to the upper limb, extending from the neck into the axilla (see figure for question 7).

Major nerves of the upper limb include the:

- **Axillary nerve** (roots C5 to C6): arises from the **posterior cord** of the brachial plexus. It winds round the **surgical neck** of the humerus, passing through the quadrangular space with the posterior circumflex humeral artery. It provides motor supply to the **deltoid** and **teres minor**, and cutaneous supply to skin over the inferior half of the deltoid
- **Radial nerve** (roots C5 to T1): arises from the **posterior cord** of the brachial plexus. It provides the major nerve supply to the **extensor muscles** of the upper limb and their cutaneous sensation to the skin. The nerve courses posterior to the brachial artery before passing between the long and medial heads of the triceps in the head of the humerus. It gives branches to the **triceps** and **brachioradialis**. At the lateral epicondyle the nerve divides into the **superficial radial nerve** and the **posterior interosseous nerve**
- **Musculocutaneous nerve** (roots C5 to C7): arises from the **lateral cord** of the brachial plexus. It passes through the **coracobrachialis** muscle before descending the arm between the **brachialis** and **biceps** (supplying all three muscles). It supplies the skin of the lateral forearm
- **Median nerve** (roots C6 to T1): arises from the medial and **lateral cords** of the brachial plexus. The nerve crosses the brachial artery medially in the mid-arm. It continues to the cubital fossa and gives off the **anterior interosseous branch**. The median nerve supplies the deep muscles of the **flexor compartment** of the forearm and skin overlying the thenar eminence
- **Ulnar nerve** (roots C8, T1): arises from the **medial cord** of the brachial plexus. It runs on the medial side of the brachial artery, and pierces the coracobrachialis in the mid-arm where it enters the posterior compartment. In the mid-forearm, the ulnar nerve is divided into two branches: the **superficial terminal branch** and the **deep terminal branch**.

The following figures illustrate their courses:

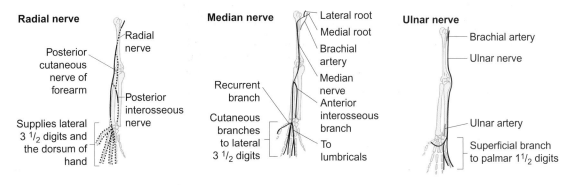

Radial nerve

Radial nerve

Posterior cutaneous nerve of forearm

Posterior interosseous nerve

Supplies lateral 3 $\frac{1}{2}$ digits and the dorsum of hand

Median nerve

Lateral root
Medial root
Brachial artery
Median nerve
Anterior interosseous branch

Recurrent branch

Cutaneous branches to lateral 3 $\frac{1}{2}$ digits

To lumbricals

Ulnar nerve

Brachial artery
Ulnar nerve
Ulnar artery
Superficial branch to palmar 1 $\frac{1}{2}$ digits

Answers

7. 1 – H, 2 – I, 3 – J, 4 – K, 5 – E, 6 – F, 7 – D, 8 – C
8. F T F T T
9. T F T F T

10. The following diagram shows a sagittal section of the axilla, illustrating its anterior wall. Label the diagram using the options provided. You may use the options once, more than once, or not at all

Options

A. Axillary artery
C. Subclavian artery
E. Pectoralis major
G. Teres major
I. Subscapularis
K. Infraspinatus
M. Axillary fascia
O. Sternocleidomastoid
Q. Serratus anterior
S. Base of axilla
U. Posterior wall of axilla
W. Lateral wall of axilla

B. Axillary vein
D. Subclavian vein
F. Pectoralis minor
H. Teres minor
J. Supraspinatus
L. Axillary sheath
N. Clavipectoral fascia
P. Latissimus dorsi
R. Apex of axilla
T. Anterior wall of axilla
V. Medial wall of axilla

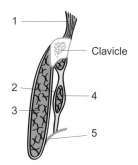

11. The following diagram shows a transverse section of the axilla and its contents. Label the diagram using the options provided in the previous question. You may use the options once, more than once, or not at all

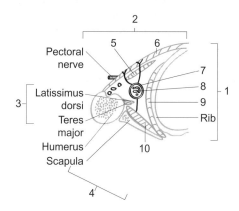

EXPLANATION: THE AXILLA AND AXILLARY LYMPH NODES

The axilla is a three-sided pyramidal space at the junction of the arm and the thorax, and consists of the:

- **Apex**: formed by the convergence of the clavicle in the anterior wall, the scapula in the posterior wall, and the first rib in the medial wall. These three bones form the entrance to the axilla, where nerves and blood vessels supplying the upper limb pass through
- **Base**: the armpit, formed by fat and skin
- **Anterior wall**: consists of the pectoral muscles, the clavicle and the clavipectoral fascia. The lateral border of the pectoralis major forms the anterior axillary fold
- **Posterior wall**: formed by the scapula and the subscapularis muscle, teres major and latissimus dorsi. The latter two combine to form the posterior axillary fold
- **Medial wall**: made up of the serratus anterior, ribs and intercostal muscles
- **Lateral wall**: consists of the floor of the intertubercular groove of the humerus.

Contents of the axilla include the:

- **Axillary artery**: the scapular anastomosis in the axilla provides collateral supply if the axillary artery is obstructed
- **Axillary vein**: drains venous return from the brachial vein of the upper limb and continues as the subclavian vein at the lateral border of the first rib
- **Branches of the brachial plexus**: a large network of nerves that extend from the neck into the axilla and supply the upper limb
- **Axillary lymph nodes**: this large group of lymph nodes in the axilla is arranged into five groups. The **anterior** and **posterior groups** lie in the anterior and posterior aspects of the medial wall of the axilla, respectively. The **lateral group** lies medial to the axillary vein. The **central group** lies within the fat of the axillary pyramidal space and the **apical group** lies in the apex of the axilla; these both receive lymph from the anterior, posterior and lateral groups and lymph is then drained into the **thoracic duct** on the left or the **right lymphatic trunks**. The **supratrochlear group** of nodes lies above the medial epicondyle of the arm and drains lymph into the lateral group. Some lymph from the radial aspect of the upper limb drains lymph into the **infraclavicular group** of nodes then the apical group.

Answers
10. 1 – O, 2 – N, 3 – E, 4 – F, 5 – M
11. 1 – V, 2 – T, 3 – W, 4 – U, 5 – B, 6 – F, 7 – E, 8 – L, 9 – A, 10 – Q, 11 – I

12. Concerning the pectoral muscles and the back muscles

a. The sternoclavicular joint is a synovial joint
b. The pectoralis major adducts the arm with the teres major
c. The serratus anterior prevents 'winging' of the scapula
d. The trapezius is supplied by CN XI
e. The latissimus dorsi helps in the adduction and medial rotation of the humerus

13. Consider the scapular muscles

a. The deltoid is supplied by the musculocutaneous nerve
b. The deltoid helps in the adduction of the humerus
c. The teres major is attached to the greater tuberosity of the humerus
d. The teres minor is attached to the greater tuberosity of the humerus
e. The subscapularis helps to medially rotate and adduct the humerus

14. Label the following diagrams from the options provided.

Options

A. Deltoid   B. Teres minor   C. Teres major
D. Supraspinatus   E. Infraspinatus   F. Subscapularis

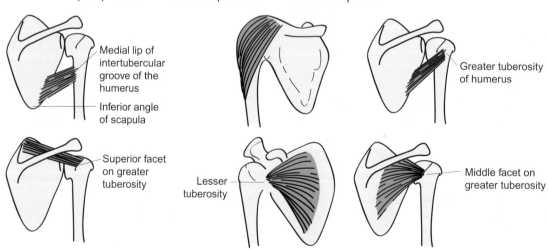

For question 15 see page 108.

EXPLANATION: THE PECTORAL AND SCAPULAR REGIONS

The upper limb is connected to the axial skeleton by the scapula and the clavicle via the **acromioclavicular** and **sternoclavicular joints** (both synovial joints), and muscular attachments, including:

The pectoral muscles:

- **Pectoralis major**: attached to the clavicle, sternum and intertubercular groove of the humerus. It adducts and medially rotates the arm. It is innervated by the pectoral nerve
- **Pectoralis minor**: attached to the third to fifth ribs proximally and the coracoid process of the scapula distally, it stabilizes the scapula. It is also innervated by the pectoral nerve
- **Serratus anterior**: innervated by the long thoracic nerve, it is attached from the first to the eighth ribs and the medial border of the scapula. It rotates and stabilizes the scapula.

The back muscles (see figures for question 15):

- **Trapezius**: attached to the superior nuchal line, the spinous processes of C7 to T12 and the spine of the scapula. It elevates, retracts, depresses and rotates the scapula, and is innervated by the accessory nerve (CN XI)
- **Latissimus dorsi**: covers the inferior half of the back, it extends, adducts and medially rotates the arm. It is attached medially to the spinous processes of the T7 to T12 and laterally to the intertubercular groove of the humerus. It is supplied by the thoracodorsal nerve
- **Levator scapulae**: supplied by the dorsal scapular nerve, it elevates the scapula, helps to fix the scapula against the trunk and to flex the neck laterally
- **Rhomboid major** and **minor**: helps to rotate the scapula. It is innervated by the dorsal scapular nerve.

The scapular muscles (see figures for question 14):

- **Deltoid**: this muscle medially rotates, abducts, extends and laterally rotates the arm. The deltoid is attached to the deltoid tuberosity of the humerus, lateral third of the clavicle and the acromion and spine of the scapula. It is supplied by the axillary nerve
- **Teres major**: supplied by the lower subscapular nerve, it adducts and medially rotates the arm. It is attached to the inferior angle of the scapula and the medial lip of the intertubercular groove
- **Teres minor**: laterally rotates and adducts the arm. It is innervated by the axillary nerve, and is attached to the lateral border of the scapula and the greater tuberosity.
- **Infraspinatus**: supplied by the suprascapular nerve, it is attached to the infraspinous fossa of the scapula and the greater tuberosity. It acts with the teres minor
- **Supraspinatus**: aids the deltoid to abduct the humerus and is innervated by the suprascapular nerve. It is attached to the supraspinous fossa of the scapula and the greater tuberosity of the humerus
- **Subscapularis**: medially rotates and adducts the arm. It is innervated by the subscapular nerves, and is attached to the subscapular fossa and the lesser tubercle.

Answers

12. T T T T T
13. F F F T T
14. 1 – C, 2 – A, 3 – B, 4 – D, 5 – F, 6 – E
15. 1 – F, 2 – D, 3 – E, 4 – C, 5 – B, 6 – A (For Question 15, see page 108).

16. Regarding the shoulder joint

a. It is a hinge joint
b. Failure of active abduction of the shoulder may indicate a torn tendon of the supraspinatus muscle
c. There is no bursa associated with the shoulder joint
d. The synovial membrane of the shoulder joint surrounds part of the tendon of the biceps
e. It is supplied by the branches of axillary artery

17. List the four rotator cuff muscles

18. List the six movements at the shoulder joint. Which muscles are involved in each of the movements?

19. The following diagram illustrates muscles in the shoulder region including the rotator cuff muscles. Label the diagram with options provided. You may use the options once or not at all

Options

A. Subscapular bursa
C. Supraspinatus
E. Teres minor

B. Subacromial bursa
D. Infraspinatus
F. Subscapularis

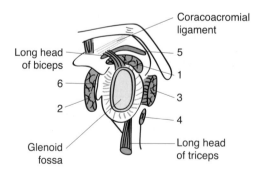

Coracoacromial ligament

Long head of biceps

5
1
6
3
2
4

Glenoid fossa

Long head of triceps

20. What makes the shoulder joint vulnerable to dislocation? Which nerve may be damaged in the process?

EXPLANATION: THE SHOULDER JOINT

The shoulder joint is a multiaxial **synovial ball and socket** joint which permits a wide range of movement. It is formed by the articulation of the head of the humerus and the glenoid fossa of the scapula. The joint is stabilized by the **rotator cuff muscles**, which include the *S*upraspinatus, *I*nfraspinatus, *T*eres minor and *S*ubscapularis (*SITS*) (**17**). The joint is also stabilized by three **glenohumeral ligaments** anteriorly, the **coracohumeral ligament** superiorly and the **coracoacromial ligament**.

The shoulder joint is enclosed by a lax fibrous capsule. The capsule is lined by a synovial membrane, which surrounds part of the tendon of the biceps brachii muscle. The joint has two large bursae. The **subscapular bursa** separates the shoulder capsule from the tendon of the subscapularis, and the **subacromial bursa** separates the shoulder capsule from the coracoacromial ligament.

The shoulder joint is supplied by the anterior and posterior **circumflex humeral branches** of the **axillary artery** and the **suprascapular branch** of the **subclavian artery**. Innervation includes the suprascapular, axillary, and lateral pectoral nerves.

Movements of the shoulder include **flexion–extension, abduction–adduction, medial–lateral rotation** and **circumduction**. Muscles involved include the:

- **Flexors**: pectoralis major, coracobrachialis and deltoid
- **Extensors**: teres major, latissimus dorsi and deltoid
- **Abductors**: supraspinatus and deltoid
- **Adductors**: pectoralis major and latissimus dorsi
- **Medial rotators**: pectoralis major, latissimus dorsi, teres major, subscapularis and deltoid
- **Lateral rotators**: infraspinatus, teres minor and deltoid (**18**).

The **supraspinatus** is also responsible for the initiation of abduction of the shoulder. Therefore failure of active abduction of the shoulder may indicate a torn tendon of the supraspinatus muscle. After the initiation of abduction, the supraspinatus then abducts with the deltoid to 90 degrees. Further elevation is achieved by rotating the scapula with the trapezius and serratus anterior.

The rotator cuff muscles stabilize the shoulder joint superiorly, anteriorly and posteriorly. However, inferiorly the shoulder is unprotected. In violent abduction, the humeral head may be dislocated inferoanteriorly (**anterior dislocation**). The axillary nerve may be damaged in the process (**20**). Posterior dislocation of the shoulder is uncommon.

Answers

16. F T F T T
17. See explanation
18. See explanation
19. 1 – C, 2 – F, 3 – D, 4 – E, 5 – B, 6 – A
20. See explanation

21. Concerning the anterior compartment of the arm

   a. The musculocutaneous nerve pierces through the coracobrachialis
   b. The biceps brachii is the primary flexor muscle of the elbow
   c. The long head of the biceps descends in the intertubercular groove of the humerus
   d. The brachialis is innervated by the ulnar nerve
   e. The biceps brachii is used when manually putting a screw into a piece of wood

22. Concerning the posterior compartment of the arm

   a. The triceps brachii is supplied by the musculocutaneous nerve
   b. The triceps brachii is used to extend the forearm
   c. The triceps brachii is attached to the olecranon of the ulna distally
   d. The absence of the triceps brachii reflex strongly indicates lesion at C6 level
   e. The ulnar nerve lies partly within this compartment

23. The following diagram shows a cross-section through the arm just above the right elbow. Label the diagram with options provided

Options

   A. Median nerve   **B.** Radial nerve
   C. Ulnar nerve   **D.** Musculocutaneous nerve
   E. Cephalic vein   **F.** Basilic vein
   G. Brachialis   **H.** Biceps brachii
   I. Triceps brachii

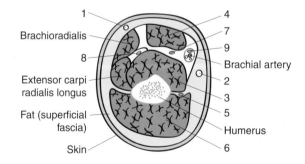

EXPLANATION: THE ARM

The **anterior compartment** of the arm consists of the:

- **Biceps brachii**: formed by two heads (long and short), with the tendon of its long head descending through the intertubercular groove of the humerus. The two bellies unite just distal to the middle of the arm. It is attached to the radial tuberosity and the deep fascia of the forearm via the bicipital aponeurosis. It flexes the forearm. It is also a powerful supinator when the forearm is flexed (for example when manually putting a screw into a piece of wood)
- **Brachialis**: lies posterior to biceps and is the main flexor of the forearm in all positions. It is attached to the distal half of the anterior surface of the humerous proximally and the coronoid process and tuberosity of the ulna distally
- **Coracobrachialis**: pierced by the musculocutaneous nerve, it functions to flex and adduct the arm. It is attached to the corocoid process of the scapula proximally and the medial surface of the humerus distally
- **Brachial artery, basilic vein, median** and **ulnar nerves**.

All muscles in the anterior compartment of the arm are innervated by the **musculocutaneous nerve**. The diagram below illustrates contents in the anterior compartment of the arm.

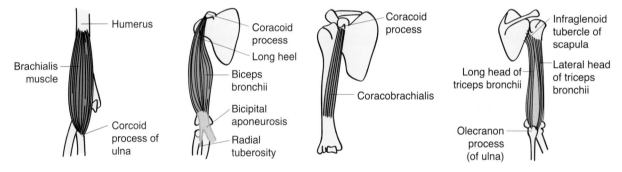

The **posterior compartment** of the arm consists of the:

- **Triceps brachii**: this muscle comprises three heads – long, medial and lateral. It is the main extensor of the forearm. It assists in the extension and adduction of the arm at the shoulder. Proximally its long head is attached to the infraglenoid tubercle of the scapula, while its medial and lateral heads are attached to the posterior surface of the humerous. The triceps attaches to the proximal end of the olecranon of the ulna distally
- **Profunda brachii artery** and the **radial nerve**.

The triceps is supplied by the **radial nerve** (C7, C8).

Answers

21. T T T F T
22. F T T F T
23. 1 – E, 2 – F, 3 – G, 4 – H, 5 – C, 6 – I, 7 – D, 8 – B, 9 – A

24. Concerning the elbow joint

a. It is a ball and socket synovial joint
b. It has a capsule that is relatively weak anteriorly and posteriorly
c. The medial collateral ligament consists of three bands
d. The lateral collateral ligament extends from the lateral epicondyle to the annular ligament
e. It may be dislocated posteriorly by a fall on the outstretched hand

25. Concerning the cubital fossa

a. It contains the ulnar nerve
b. It contains the median nerve
c. It contains the brachial artery
d. The brachialis forms the lateral border
e. The pronator teres forms the medial border

26. What movements occur at the elbow joint?

27. The following diagram illustrates an anteroposterior view of the cubital fossa. Label it from the options provided. You may use the options once, more than once, or not at all

A. Ulnar nerve
B. Median nerve
C. Radial nerve
D. Profunda brachii artery
E. Brachial artery
F. Ulnar collateral artery
G. Triceps brachii
H. Pronator teres
I. Supinator
J. Bicipital aponeurosis
K. Flexor carpi radialis
L. Flexor digitorum superficialis
M. Palmaris longus
N. Interosseous membrane
O. Medial collateral ligament
P. Lateral collateral ligament
Q. Annular ligament
R. Tendon of biceps brachii
S. Brachialis
T. Brachioradialis

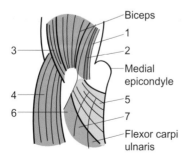
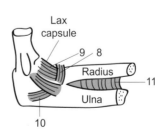

EXPLANATION: THE ELBOW JOINT AND THE CUBITAL FOSSA

The elbow joint is a **synovial hinge joint**. It has three articulations the:

- **Humero-ulnar**: between the trochlea of the humerus and the trochlear notch of the ulna
- **Humero-radial**: between the capitulum and radial head
- **Superior radio-ulnar**: between the side of the radial head and the radial notch of the ulna.

Movements at the elbow include **flexion–extension** and **pronation–supination (26)**. Posterior dislocation of the elbow may be caused by the indirect trauma of a fall on the outstretched hand.

The **capsule** of the joint is relatively weak anteriorly and posteriorly to permit flexion and extension. The sides of the capsule are strengthened by the medial and collateral ligaments:

- **Medial collateral ligament**: consists of the anterior, posterior and middle bands. They extend from the medial epicondyle and the olecranon of the humerus to the coronoid process of the ulna. The ulnar nerve can be found medial to this ligament
- **Lateral collateral ligament**: is attached proximally to the lateral epicondyle and distally to the **annular ligament**, which loops around the radial head in the upper margin but is free in the lower margin to allow rotation of the radius.

The **cubital fossa** is a triangular area on the anterior surface of the elbow. The borders of the fossa are as follows:

- **Superior border**: an imaginary line connecting the **epicondyles** of the humerus
- **Medial border**: lateral border of the **pronator teres**
- **Lateral border**: medial border of the **brachioradialis**. The brachioradialis is the most superficial muscle on the radial side of the forearm. It arises from the lateral epicondyle of the humerus and is supplied by the radial nerve. Its main action is to flex the forearm
- **Floor**: the **brachialis** and the **supinator**. The supinator is also the prime muscle in supination of the forearm and hand. It arises from the lateral epicondyle of the humerus and is supplied by the radial nerve.
- **Roof: superficial fascia**.

The cubital fossa contains (medial to lateral) the **median nerve**, the **brachial artery**, the **brachial veins**, the **biceps tendon**.

Answers

24. F T T T T
25. F T T F T
26. See explanation
27. 1 – B, 2 – E, 3 – S, 4 – T, 5 – J, 6 – H, 7 – K, 8 – Q, 9 – P, 10 – O, 11 – N

28. Concerning the forearm

a. The radial nerve innervates the flexor carpi radialis
b. The flexor carpi ulnaris is innervated by the ulnar nerve
c. Tendons of the palmaris longus pass through the flexor retinaculum
d. The flexor carpi radialis is attached to the medial epicondyle of the humerus proximally
e. The ulnar nerve lies between the bellies of the three heads of the flexor digitorum superficialis

29. Concerning the forearm

a. Tendons of the flexor digitorum profundus pass anterior to tendons of the flexor digitorum superficialis
b. The flexor pollicis longus flexes the distal phalanx of the thumb
c. The flexor pollicis longus lies medial to the flexor digitorum profundus
d. The flexor digitorum superficialis flexes the distal phalanges of the medial four digits
e. The pronator teres is the prime pronator of the forearm

30. The following diagrams show contents of the anterior compartment of the forearm. Label them with the options provided

Options

A. Flexor digitorum superficialis
C. Flexor pollicis longus
E. Flexor carpi radialis
G. Pronator teres
I. Brachioradialis

B. Flexor digitorum profundus
D. Flexor carpi ulnaris
F. Pronator quadratus
H. Palmaris longus

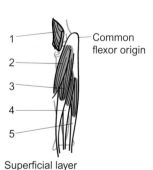

1 ⎯⎯ Common flexor origin
2 ⎯
3 ⎯
4 ⎯
5 ⎯

Superficial layer

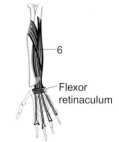

6

Flexor retinaculum

Intermediate layer

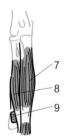

7
8
9

Deep layer

EXPLANATION: THE FOREARM: THE ANTERIOR COMPARTMENT

Muscles in the anterior (**flexor**) compartment of the forearm can be divided into three layers the (see figures in question 30):

- **Superficial layer** (from lateral to medial):

  - **Pronator teres**: this is a pronator of the forearm and a flexor of the elbow joint
  - **Flexor carpi radialis**: this muscle flexes and abducts the hand. Its tendon lies medial to the radial pulse
  - **Palmaris longus**: tendons of this muscle pass superficial to the flexor retinaculum. The median nerve lies lateral to it at the wrist. It tenses the palmar aponeurosis and assists in the flexion of the hand
  - **Flexor carpi ulnaris**: the ulnar nerve and artery lie on the lateral side of the tendon of the flexor carpi ulnaris at the wrist. It flexes and adducts the hand.

- **Intermediate layer**:

  - **Flexor digitorum superficialis**: the median nerve and ulnar artery pass between the three heads of this muscle. It gives rise to four tendons at the wrist which pass deep to the flexor retinaculum before attaching to the bodies of the middle phalanges of the medial four digits. It flexes the metacarpophalangeal and proximal interphalangeal joints of the medial four digits, and the wrist joint

- **Deep layer**:

  - **Flexor pollicis longus**: The tendon of this muscle passes through the flexor retinaculum and flexes the distal phalanx of the thumb. It is attached to the anterior surface of the radius proximally and the base of the distal phalanx of the thumb distally.
  - **Flexor digitorum profundus**: Originated from the anterior surface of the ulna, its four tendons pass posterior to the flexor digitorum superficialis and through the flexor retinaculum. The four tendons then attach to the bases of the distal phalanges of the medial four digits. It functions to flex the distal interphalangeal joints of the medial four digits
  - **Pronator quadratus**: the prime pronator of the forearm. It is the deepest muscle of the layer. This 'band' of muscle crosses between the distal surfaces of the ulna and radius.

All muscles in the superficial layer and part of the flexor digitorum superficialis arise from the **common flexor origin** at the medial epicondyle of the humerus. All muscles in the compartment are supplied by the **median nerve**, except for the flexor carpi ulnaris and the medial half of the flexor digitorum profundus (**ulnar nerve**). Muscles in the anterior compartment are supplied by the radial artery, the ulnar artery, and its anterior interosseous branch.

Answers

28. F T F T F
29. F T F F F
30. 1 – I, 2 – G, 3 – E, 4 – H, 5 – D, 6 – A, 7 – B, 8 – C, 9 – F

31. Concerning the forearm

a. The extensor carpi radialis brevis extends and abducts the hand at the wrist joint
b. The extensor carpi radialis longus is innervated by the radial nerve
c. The ulnar nerve innervates the extensor carpi ulnaris
d. The common extensor origin is located at the medial epicondyle of the humerus
e. The extensor carpi radialis brevis inserts into the trapezium bone

32. Concerning the forearm

a. The abductor pollicis longus passes superficial to the extensor retinaculum
b. The extensor pollicis brevis extends the proximal phalanx of the thumb
c. The extensor indicis extends the index finger and the hand
d. The extensor pollicis longus is innervated by the posterior interosseous nerve
e. There is an interosseous membrane between the radius and ulna

33. Name the three muscles/tendons that form the boundaries of the anatomical snuffbox

34. The following diagrams show contents of the posterior compartment of the forearm. Label them with the options provided. You may use the options once, more than once, or not at all

A. Extensor digiti minimi
B. Extensor carpi radialis longus
C. Extensor carpi radialis brevis
D. Extensor carpi ulnaris
E. Extensor digitorum
F. Abductor pollicis longus
G. Abductor pollicis brevis
H. Extensor indicis
I. Extensor pollicis longus

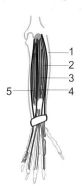

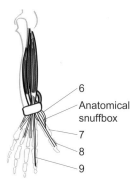

Anatomical snuffbox

EXPLANATION: THE FOREARM: THE POSTERIOR COMPARTMENT

Muscles in the posterior (extensor) compartment of the forearm are divided into two layers:

- **Superficial layer** (see first panel of question 34):

  - **Extensor carpi radialis longus**: partly posterior to the brachioradialis, it abducts and extends the hand
  - **Extensor carpi radialis brevis**: covering the extensor carpi radialis longus, it abducts and extends the hand with the extensor carpi radialis longus
  - **Extensor digitorum**: its four tendons extend the medial four digits
  - **Extensor digiti minimi**: this muscle joins the tendon of the extensor digitorum and the fifth digit. It extends the proximal phalanx of the fifth digit
  - **Extensor carpi ulnaris**: extends and adducts hand.

All muscles in this layer arise from the **common extensor origin** on the lateral epicondyle of the humerus except for the extensor carpi radialis longus which is attached to the lateral supracondylar ridge. All their tendons pass through the **extensor retinaculum**.

- **Deep layer** (see second panel of question 34):

  - **Abductor pollicis longus**: this muscle lies just distal to the supinator. It abducts and extends the thumb at the carpometacarpal joint
  - **Extensor pollicis brevis**: lying distal to and partly covered by the abductor pollicis longus, it extends the proximal phalanx of the thumb
  - **Extensor pollicis longus**: this muscle extends the distal phalanx of the thumb at the metacarpophalangeal and interphalangeal joints
  - **Extensor indicis**: lying medial to the extensor pollicis longus, it extends the index finger and the hand.

Tendons of the **extensor pollicis brevis** and **abductor pollicis longus** form the lateral boundary of the **anatomical snuffbox**, while the **extensor pollicis longus** forms its medial border **(33)**. Muscles in the posterior compartment of the forearm are supplied by the **posterior interosseous artery**. They are all innervated by the **posterior interosseous nerve** of the radial nerve.

Answers

31. T T F F F
32. F T T T T
33. See explanation
34. 1 – B, 2 – C, 3 – E, 4 – A, 5 – D, 6 – F, 7 – G, 8 – I, 9 – H

35. Regarding joints in the wrist and hand

a. The wrist joint is a plane synovial joint
b. Interphalangeal joints are synovial hinge joints
c. The saddle shape of the carpometacarpal joint of the thumb permits rotation of the thumb
d. The flexor retinaculum flexes the wrist
e. The proximal part of the palmar aponeurosis is continuous with the flexor retinaculum

36. What are the functions of the interossei muscles? What are the movements of the thumb?

37. Regarding muscles of the hand

a. There are five lumbricals in each hand
b. The abductor pollicis brevis opposes the thumb
c. The hypothenar muscles concern movements of the fifth digit
d. The abductor digiti minimi is innervated by the ulnar nerve
e. The interosseous muscles are innervated by the median nerve

38. The figure shows the superficial muscles of the hand. Label them with the options provided. You may use the options once, more than once, or not at all

A. Flexor digiti minimi
B. Flexor pollicis brevis
C. Opponens pollicis
D. Opponens digiti minimi
E. Abductor pollicis brevis
F. Abductor digiti minimi

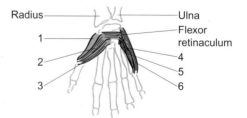

EXPLANATION: THE WRIST AND THE HAND

The **palmar aponeurosis** is a triangular deep fascia of the hand overlying the flexor tendons of the palm. Its proximal end is continuous with the **flexor retinaculum**, a fibrous band overlying the carpal bones, and prevents tendons from springing away from the wrist. The space between the flexor retinaculum and the carpal bones forms the carpal tunnel. Compression of the tunnel causes **carpal tunnel syndrome** (see page 107).

The **wrist joint** is a **condyloid synovial joint**, where the distal end of the radius articulates with the scaphoid, lunate and triquetral. Enclosed by a fibrous capsule and strengthened by the radial and ulnar collateral ligaments, movements at the wrist joint include **flexion–extension** and **abduction–adduction**. **Intercarpal joints** (plane synovial joints) are between the carpal bones. The **carpometacarpal joint of the thumb** is a saddle-type of synovial joint between the trapezium and the first metacarpal, permitting angular movements in any plane. The **metacarpophalangeal joints** are synovial condyloid joints, while the **interphalangeal joints** are synovial hinge joints.

Muscles of the hand can be divided into four groups:

- **Thenar muscles**: these include the **abductor pollicis brevis**, **flexor pollicis brevis** and **opponens pollicis**. The **adductor pollicis** is sometimes included in this muscle group but it is not part of the thenar eminence. They are responsible for **opposition** of the thumb. All are innervated by the **median nerve**, except for adductor pollicis which is supplied by the **ulnar nerve**
- **Hypothenar muscles**: these are the **abductor digiti minimi** (adducts fifth digit), **flexor digiti mini brevis** (flexes proximal phalanx of fifth digit) and **opponens digiti minimi** (brings the fifth digit into opposition with thumb). They are concerned with the movements of the **fifth digit**. They are all supplied by the **ulnar nerve** (see figure for question 38)
- **Lumbricals**: these **four** muscles flex the digits at the metacarpophalangeal joints and extend the interphalangeal joints. Arising from tendons of the flexor digitorum profundus, they insert into the radial aspect of each of the proximal phalanges and into the extensor expansions. The lateral two lumbricals are supplied by the **median nerve**, and the medial two lumbricals are supplied by the **ulnar nerve**
- **Interosseous muscles**: these seven interossei are located between the metacarpal bones, arranged into two layers: three **palmar** and four **dorsal**. The **p**almar interossei **ad**duct (**PAD**) the digits while the **d**orsal interossei **ab**duct (**DAB**) the digits. Supplied by the **ulnar nerve**, they help the lumbricals in the flexion of the metacarpophalangeal joints and extension of the interphalangeal joints (**36**) (see figure for question 38).

Movements of the thumb include **extension** (point thumb laterally), **opposition** (thumb to tip of fifth digit), **abduction** (point up), **adduction** (thumb in palm) and **flexion** (thumb across palm) (**36**).

Answers

35. F T F F T
36. See explanation
37. F T T T F
38. 1 – C, 2 – E, 3 – B, 4 – D, 5 – A, 6 – F

39. Case study

A 50-year-old gardener visits her GP complaining of a severe 'pins and needles' sensation in the palmar surface of her lateral three-and-a-half digits of her right hand. She has difficulties in using her tools and cannot do her job properly as she used to. The pain is getting worse and sometimes it even wakes her up at night. On examination the GP noticed that the thenar muscles on the patient's right hand were considerably weaker than those on the left.

A. What is the likely diagnosis?
B. Explain how these signs arose.
C. How can the condition be treated?

40. Case study

An X-ray shows that a patient has suffered from a fracture of the surgical neck of the humerus.

A. Which nerve is most likely to have been damaged?
B. Which muscle is most likely to have been weakened?

41. Case study

A patient has been involved in a trauma of his right upper limb. On examination there is loss of abduction and adduction of the digits, loss of hypothenar muscle functions as well as that of the medial two lumbricals. Sensation of the lateral one-and-a-half digits is lost. Wrist flexion is also weakened and upon flexion of the wrist the hand is deviated to the radial side.

A. Which nerve is likely to have been damaged in this patient?
B. At what level is the damage likely to be?

42. Case study

After birth a baby presents with her right arm hanging down by the side. Her arm is internally rotated, her forearm is pronated and the hand is flexed.

A. What is the likely diagnosis?
B. What roots are likely to have been involved?

GP, general practitioner

EXPLANATION: CLINICAL SCENARIOS

39.

This lady presents with signs of **carpal tunnel syndrome**. The carpal tunnel is a space bounded by the flexor retinaculum anteriorly and the carpal bones posteriorly. It contains nine tendons: four tendons of the flexor digitorum superficialis, four tendons of the flexor digitorum profundus, and the tendon of the flexor pollicis longus. The **median nerve** is also present within the tunnel. The syndrome results from the compression of the median nerve within the tunnel. The median nerve supplies sensation in the lateral three-and-a-half digits and hence the 'pins and needles' sensation experienced by the patient. The thenar muscles and the lateral two lumbrical muscles are also supplied by the median nerve; the function of the first to third digits may diminish, leading to the patient having difficulties in performing her job properly. There may also be wasting or atrophy of the thenar muscles. The condition can be relieved by a **carpal tunnel release** operation, where the flexor retinaculum is divided surgically.

40.

Fracture of the surgical neck may damage the **axillary nerve**, hence the patient may present with a weakened **deltoid muscle**.

41.

Damage to the **ulnar nerve** presents with the loss of abduction and adduction ability of the digits, loss of function of the hypothenar muscles and the medial two lumbricals. The sensation of digits five and half that of four may be diminished. A claw hand may be seen on examination. Wrist flexion is only weakened and deviated radially when flexed if the damage to the ulnar nerve is at the **elbow** (not the wrist). This is because the flexor carpi ulnaris is affected.

42.

The baby presents with classical signs of **Erb–Duchenne paralysis (Erb's palsy)**, and these signs have been termed the 'waiter's tip' position. During birth **C5** and **C6** roots may be injured by the excessive downward traction on the upper limb, resulting in paralysis of the deltoid, brachialis, biceps, and short muscles of the shoulder.

Answers

39. See explanation
40. See explanation
41. See explanation
42. See explanation

15. Label the following diagram from the options provided. You may use the options once, more than once, or not at all

Options

A. Rhomboid major   **B.** Rhomboid minor
C. Latissimus dorsi   **D.** Deltoid
E. Infraspinatus   **F.** Trapezius

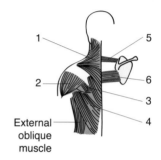

6 THE LOWER LIMB

1. Concerning bones of the lower limb

a. The femoral head articulates with the acetabulum
b. The lesser trochanter projects anterolaterally from the junction of the neck and the femoral shaft
c. The linea aspera is the rough ridge of bone in the middle of the posterior surface of the femur
d. Both the tibia and the fibula are weight-bearing bones of the lower limb
e. The distal end of the tibia forms the medial malleolus

2. Concerning bones of the foot

a. The talus articulates with the tibia
b. The calcaneus forms the heel of the foot
c. The cuboid is the largest bone in the foot
d. The bases of the metatarsals articulate with the cuneiform and cuboid bones
e. There are 12 phalanges

3. The following diagrams show bones of the lower limb. Label them with the options provided

Options

A. Spiral line	**B.** Epiphyseal line	**C.** Shaft of femur
D. Shaft of tibia	**E.** Shaft of fibula	**F.** Linea aspera
G. Head of femur	**H.** Greater trochanter	**I.** Lesser trochanter
J. Trochanteric fossa	**K.** Medial malleolus	**L.** Lateral malleolus
M. Medial condyle of femur	**N.** Lateral condyle of femur	**O.** Inter-trochanter-line

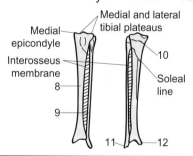

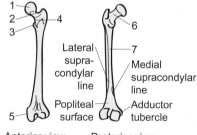

4. The following diagram shows bones of the foot. Label them with the options provided

Options

A. Cuboid	**B.** Calcaneus	**C.** Cuneiform	**D.** Navicular
E. Metatarsus	**F.** Tarsus	**G.** Phalanges	**H.** Talus

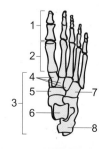

EXPLANATION: BONES OF THE LOWER LIMB

Bones of the lower limb consist of the:

- **Femur**: is the longest bone in the body. The round femoral **head** projects superomedially and articulates with the **acetabulum** of the hip. The central depression on the head is called the **fovea**. The femoral **neck** connects the head to the **shaft** of the femur. The **greater trochanter** of the femur is a large projection from the junction of the neck and the shaft. The **lesser trochanter** projects posteromedially between the neck and the shaft of the femur. The shaft of the femur has an anterior convexity. The rough ridge of bone in the middle of its posterior surface is termed the **linea aspera**. This consists of a medial and a lateral lip, which diverge to form the medial and lateral **supracondylar lines** inferiorly. The medial supracondylar line terminates at the **adductor tubercle**. The distal end of the femur consists of the medial and lateral **femoral condyles**. They project posteriorly and are separated by a deep, U-shaped **intercondylar notch** (see figure for question 3)
- **Tibia**: is a large weight-bearing bone articulating with the condyles of the femur superiorly, with its flattened medial and lateral **tibial plateaus**. The tibia has an **intercondylar eminence** superiorly which fits into the intercondylar notch of the femur. At the superoanterior aspect of the shaft of the tibia is the **tibial tuberosity**. The superoposterior surface of the shaft has an oblique line, termed the soleal line. The distal end of the tibia projects medially and inferiorly as the **medial malleolus**. The distal tibia articulates with the talus
- **Fibula**: is a slender bone lying posterolateral to the tibia. It is attached to the tibia at the superior (synovial) and inferior (fibrous) tibiofibular joints. Between the two bones is an **interosseous membrane**. The fibula has **no** function in weight bearing. It does not form part of the knee joint. The **neck** of the fibula separates the **head** from the **shaft**. The distal end of the fibula forms the **lateral malleolus**
- **Patella**: is a triangular sesamoid bone. Its apex is directed inferiorly, and is enclosed by the tendon of the quadriceps femoris. The ligamentum patellae attaches the patella to the tibial tuberosity

Bones of the foot (see figure for question 4) consist of the tarsus, metatarsus and phalanges:

- **Tarsus**: consists of seven tarsal bones:
  - **Talus**: The pulley-shaped superior surface (the trochlea) articulates with the inferior surface of the tibia. It rests on the anterior two-thirds of calcaneous and also articulates with the navicular anteriorly
  - **Calcaneus**: is rectangular in shape and is the largest and strongest bone of the foot. It articulates with the talus superiorly and the cuboid anteriorly
  - **Navicular**: articulates with the three cuneiforms anteriorly and the talus posteriorly
  - **Cuboid**: is the most lateral bone in the distal row of the tarsus. It articulates with the fourth and fifth metatarsal bones anteriorly and the calcaneous posteriorly
  - **Cuneiform Bones**: the medial, intermediate and lateral cuneiforms articulate with the base of its appropriate metatarsal anteriorly and the navicular posteriorly
- **Metatarsus**: consists of five metatarsal bones, with their heads articulate with the proximal phalanges and their bases with the cuneiform and cuboid bones
- **Phalanges**: There are 14 phalanges in the foot: 2 in the first digit and three each in the other four digits

Answers
1. T F T F T
2. T T F T F
3. 1 – G, 2 – H, 3 – I, 4 – O, 5 – M, 6 – J, 7 – F, 8 – D, 9 – E, 10 – B, 11 – L, 12 – K
4. 1 – G, 2 – E, 3 – F, 4 – C, 5 – D, 6 – H, 7 – A, 8 – B

5. Concerning the arterial system of the lower limb

    a. The femoral sheath encloses the femoral artery, femoral vein and femoral nerve
    b. The circumflex femoral arteries supply the anterior compartment of the thigh
    c. The anterior tibial artery descends between the extensor hallucis longus and tibialis anterior muscles
    d. The anterior tibial artery supplies the lateral compartment of the leg
    e. The posteior tibial artery supplies the flexors of the leg

6. Name four clinically important sites in the lower limb where arterial pulses can be felt in the lower limb

7. The following diagram illustrates the arterial system of the lower limb. Please label the diagram with the options provided

Options

    A. Femoral artery
    B. Circumflex femoral artery
    C. Popliteal artery
    D. Profunda femoris artery
    E. Posterior tibial artery
    F. Anterior tibial artery
    G. External iliac artery
    H. Medial plantar artery
    I. Lateral plantar artery

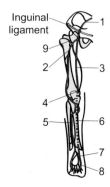

EXPLANATION: ARTERIAL SYSTEM OF THE LOWER LIMB

The **femoral artery** is a continuation of the **external iliac artery**. It enters the lower limb behind the inguinal ligament at the mid-inguinal point. Its first 2–3 cm lies against the head of the femur, where the **femoral pulse** can be felt (6). Together with the femoral vein, it is enclosed by the **femoral sheath** in the groin, while the femoral nerve lies lateral to the sheath. The artery descends through the **sub-sartorial canal** before penetrating the **adductor hiatus** of the adductor magnus muscle to become the **popliteal artery**. The largest branch of the artery is the **profunda femoris artery**. It is the chief artery to the medial and posterior compartments of the thigh, arising from the lateral side of the femoral artery just inferior to the inguinal ligament. It also gives rise to the medial and lateral **circumflex femoral branches**, supplying the head and neck of the femur. The profunda descends deep to adductor longus in the thigh and gives rise to four **perforating arteries**.

The **popliteal artery pulse** can be felt in the popliteal fossa with difficulty as it is the deepest structure in the region (6). The **popliteal artery** descends first on the posterior surface of the femur before passing under the fibrous arch of the soleus, where it bifurcates into the **anterior** and **posterior tibial arteries**. The **posterior tibial pulse** can be felt just posterior to the medial malleous (6).

The **anterior tibial artery** descends anteriorly through the interosseous membrane then down its anterior surface, between the extensor hallucis longus and tibialis anterior muscles (accompanied by the deep peroneal fibular nerve). It supplies the **anterior compartment** of the leg. The anterior tibial artery ends at the ankle joint and becomes the **dorsalis pedis artery**, where a pulse can be felt (6).

The **posterior tibial artery** supplies the flexor compartment of the leg. It gives rise to the **peroneal artery** which supplies the lateral compartment of the leg. Inferior to the medial malleous the artery descends between the tendons of the flexor hallucis longus and flexor digitorum longus. At the ankle the posterior tibial runs posterior to the medial malleolus and is divided into the medial and lateral **plantar arteries**.

The lateral plantar artery runs in the lateral part of the sole between the flexor accessories and the flexor digitorum brevis. Here, it is divided into the superficial and deep branches. The latter continues as the **deep plantar arch**, giving rise to the plantar metatarsal branches which, together with the medial plantar artery, supply the toes.

Answers

5. F T T F T

6. See explanation

7. 1 – G, 2 – D, 3 – A, 4 – C, 5 – F, 6 – E, 7 – I, 8 – H, 9 – B

8. Concerning the venous system of the lower limb

a. The great saphenous vein is the longest vein in the body
b. The small saphenous vein joins the femoral vein
c. The superficial and deep venous systems communicate via perforating veins
d. The soleal plexus is situated in the popliteal fossa
e. Deep veins in the calf do not have valves

9. Concerning the lymphatics of the lower limb

a. Deep inguinal nodes are found medial to the femoral vein
b. Cloquet's node can be found in the inguinal canal
c. Popliteal nodes are deep lymph nodes
d. The superficial inguinal group lies around the saphenous opening
e. The horizontal limb of the superficial inguinal group receives superficial lymph from the lower back and the external genitalia

10. The following diagram shows the venous system of the lower limb. Please label the diagram with the options provided

Options

A. Great saphenous vein   **B.** Small saphenous vein
C. Femoral vein   **D.** Popliteal vein
E. Perforating veins

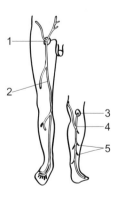

EXPLANATION: VENOUS AND LYMPHATIC SYSTEMS OF THE LOWER LIMB

The venous system of the lower limb can be divided into the **superficial** and the **deep** venous systems. The two sets communicate via **perforating veins**. All veins in the lower limbs have **valves**. They direct blood flow from the superficial to the deep venous systems, where blood is pumped upwards by muscular contractions of the calf muscles.

The longest vein in the body, the **great saphenous vein** arises from the medial side of the dorsal network of veins and ascends to the groin through the subcutaneous connective tissue layer. It courses anterior to the medial malleolus of the tibia, accompanied by the saphenous nerve. It then ascends to the medial aspect of the knee, lying superficial to the medial condyle. The vein then passes superolaterally, piercing the deep fascia to enter the **saphenous opening** where blood is drained into the **femoral vein**.

The **small saphenous vein** arises from the lateral end of the dorsal venous network of the foot, coursing posterior to the lateral malleolus. It perforates the deep fascia in the popliteal fossa to join the **popliteal vein**.

The deep veins of the lower limb are the venae comitantes of the anterior and posterior tibial arteries in the calf. They join together to form the **popliteal** and **femoral veins**. They form the **soleal plexus** in the calf, enabling the upward flow of blood against gravitational forces by muscular contractions.

The **lymph nodes** in the groin are also divided into superficial and deep groups. The **deep inguinal nodes** are found medial to the femoral vein. One of the most commonly found lymph nodes in the **femoral canal** is known as **Cloquet's node**. The deep lymph nodes receive lymph from the fascia lata of the lower limb, the skin and superficial tissues of the heel and the lateral part of the foot through the **popliteal nodes** behind the knee.

The **superficial inguinal group** lies within the superficial fascia, around the saphenous opening. They are arranged in the shape of a 'T'. The **vertical limb** receives lymph from the majority of the superficial tissues of the lower limb, whereas the **horizontal limb** receives lymph from the lower back, the external genitalia and the anterior abdominal wall. They drain lymph to the deep inguinal lymph nodes before passing to the abdomen.

Answers

8. T F T F F
9. T F T T T
10. 1 – C, 2 – A, 3 – D, 4 – B, 5 – E

11. Concerning the lumbar plexus

    a. Lies embedded within the psoas major
    b. Arises from the dorsal rami of L2 to L5
    c. Gives rise to the sciatic nerve
    d. Receives post-ganglionic nerves from the sympathetic trunk
    e. Gives rise to the obturator nerve

12. Concerning the femoral nerve

    a. Is a branch of the lumbar plexus
    b. Gives rise to the lateral femoral cutaneous nerve in the thigh
    c. Passes lateral to the psoas major muscle
    d. Lies medial to the femoral sheath
    e. Supplies the quadratus femoris muscle

13. Concerning the obturator nerve

    a. Arises from L2 to L4
    b. Has a saphenous nerve branch supplying skin over the medial aspect of the leg and foot
    c. Passes medial to the internal iliac vessels
    d. Leaves the pelvis through the obturator foramen
    e. Supplies the extensor muscles of the knee

14. Regarding the femoral nerve, obturator nerve and the lateral cutaneous nerve of the thigh, which one(s) pass under the inguinal ligament?

EXPLANATION: NERVOUS SYSTEM OF THE LOWER LIMB (i)

The **lumbar plexus** is a network of nerves formed by the anterior rami of **L1** to **L4** within the psoas major muscle. All lumbar anterior rami receive post-ganglionic sympathetic nerves from the sympathetic trunk, with L1 and L2 sending white rami communicantes to the sympathetic trunk. It has the following branches:

- **Femoral nerve**: arises from the posterior divisions of the lumbar plexus (L2 to L4). This nerve pierces through **psoas major** and then descends posterolaterally through the iliac fossa to the midpoint of the inguinal ligament, where it lies on the iliacus muscle. The nerve continues at the lateral aspect of the femoral sheath before it divides into various terminal branches just distal to the femoral triangle. One of these is the **saphenous nerve**, which supplies the skin over the medial aspect of the leg and foot. In the thigh, the femoral nerve supplies the extensor muscles of the knee
- **Obturator nerve**: originates from the anterior divisions of the lumbar plexus (L2 to L4). It descends through the medial border of the psoas major muscle. It then pierces the psoas fascia, passing laterally to the internal iliac vessels and ureter, and leaves the pelvis through the obturator foramen. The obturator nerve is then divided into the anterior and posterior divisions, supplying the adductor muscles of the thigh
- **Lateral cutaneous nerve of the thigh**: arises from L2 and L3. It passes through psoas major and crosses the iliac fossa over the iliacus. The nerve runs under the lateral part of the inguinal ligament to supply the skin on the anterolateral surface of the thigh
- **Intra-abdominal branches.**

The following diagram illustrates the femoral nerve and the obturator nerve.

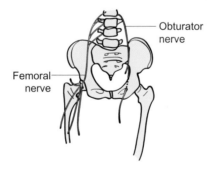

Femoral nerve

Obturator nerve

Answers

11. T F F T T
12. T F T F F
13. T F F T F
14. Obturator nerve, lateral cutaneous nerve of the thigh

15. Concerning the sacral plexus and its branches

    a. The superior gluteal nerve arises from L4-S1

    b. The posterior cutaneous nerve of the thigh supplies skin of the scrotum, buttock and back of the thigh

    c. The perforating cutaneous nerve passes through the greater sciatic foramen

    d. The inferior gluteal nerve passes through the lesser sciatic foramen

    e. The sacral plexus originates from the anterior primary rami of L2 to S4

16. Concerning the sciatic nerve

    a. It is the longest nerve in the body

    b. It passes through the greater sciatic foramen

    c. It innervates the gluteal maximus muscle

    d. Its tibial branch supplies the lateral compartment of the leg

    e. Its common peroneal branch courses over the neck of the fibula

17. The following diagram shows a posterior view of nerves in the lower limb. Label them with the options provided

Options

    A. Medial plantar nerve

    B. Lateral plantar nerve

    C. Sciatic nerve

    D. Tibial nerve

    E. Common peroneal nerve

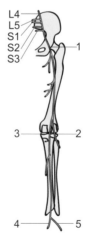

EXPLANATION: NERVOUS SYSTEM OF THE LOWER LIMB (ii)

The **sacral plexus** originates from the anterior primary rami of **L4** to **S4**. Its major nerves leave the pelvis through the **greater sciatic foramen**. Branches of the sacral plexus include the:

- **Superior gluteal nerve:** (L4-S1) supplies the gluteus medius and minimus
- **Inferior gluteal nerve:** (L5-S2) supplies the gluteus maximus
- **Posterior cutaneous nerve of the thigh:** (S2–3) supplies the skin of the scrotum, buttock and back of the thigh down to the knee
- **Perforating cutaneous nerve:** (S2–3) supplies the skin of the buttock
- **Pudendal nerve:** (S2–4) passes through the greater sciatic foramen and back into the perineum through the lesser sciatic foramen. (see page 49).
- **Sciatic nerve:** is the largest nerve in the body, formed by the anterior rami of **L4** to **S3**. It passes through the greater sciatic foramen, coursing just inferior to the piriformis muscle. It then enters the gluteal region inferior to the piriformis. It then courses inferolaterally deep to the gluteus maximus muscle before passing over the obturator internus, quadratus femoris and adductor magnus, descending in the midline. However, the sciatic nerve has **no branches** in the gluteal region. The sciatic nerve has two major branches:
  - **Tibial nerve:** traverses the popliteal fossa over the popliteal vessels before passing deep to the fibrous arch of the soleus. It courses deep to the superficial stratum of muscles at the back of the leg, grooving between the flexor digitorum longus and flexor hallucis longus. The nerve runs behind the medial malleolus deep to the flexor retinaculum, where it divides into the **medial** and **lateral plantar nerves**. The tibial nerve supplies the posterior compartments of the thigh and leg
  - **Common peroneal nerve:** descends along the medial border of the biceps tendon. It then courses over the neck of fibula before dividing into the **superficial** and **deep peroneal nerves**. The former supplies the lateral compartment of the leg, while the latter runs with the anterior tibial vessels over the interosseous membrane and supplies all the muscles of the anterior compartment.

Answers

15. T T T F F
16. T T F F T
17. 1 – C, 2 – E, 3 – D, 4 – B, 5 – A

18. Concerning muscles in the gluteal region

    a. The gluteal region is bounded by the gluteus minimus inferiorly
    b. The obturator internus attaches to the greater trochanter distally
    c. The gluteus maximus is essential when walking
    d. The gluteus medius is innervated by the inferior gluteal nerve
    e. The piriformis attaches to the greater trochanter

19. What are the movements of the hip joint?

20. The following diagrams show muscles in the gluteal region. Label them with the options provided

Options

    A. Gluteus maximus   **B.** Gluteus medius
    C. Gluteus minimus   **D.** Superior gemellus
    E. Inferior gemellus   **F.** Quadratus femoris
    G. Piriformis   **H.** Obturator internus

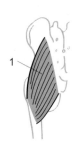 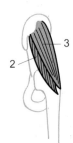 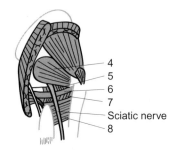

21. Concerning the hip joint

    a. It is a ball and socket joint
    b. The acetabular notch increases the depth of the acetabulum
    c. The joint has a fibrous capsule
    d. It is supplied by the circumflex femoral artery
    e. The round ligament of the head of the femur passes to the fovea of the femoral head

22. Name the ligaments that stabilize the hip joint

EXPLANATION: THE GLUTEAL REGION AND THE HIP JOINT

The **gluteal region** is the great mass of muscle that makes up the buttock. It is bounded by the iliac crest superiorly and the gluteus maximus inferiorly. It consists of the greater and lesser sciatic foramina where structures pass from the pelvis to the gluteal region. The foramina are formed by the pelvis and sacrotuberous and sacrospinous ligaments. Contents of the gluteal region include the **sciatic nerve** and its branches, the **superior** and **inferior gluteal arteries** and **muscles**.

The **gluteus maximus** is supplied by the inferior gluteal nerve. It is the chief **extensor of the thigh** and steadies it. It also assists in **lateral rotation** of the **thigh** and is used only when force is required. The **gluteus medius** and **gluteus minimus** lie deep to gluteus maximus on the external surface of the ilium. They are powerful **abductors** of the **hip joint** and hence are essential during locomotion. The gluteus minimus also helps to **rotate** the **thigh medially**. Both are supplied by the superior gluteal nerve. The **piriformis** originates from the front of the sacrum. It leaves the pelvis through the greater sciatic foramen and attaches to the greater trochanter. Together with the obturator internus and the gemelli muscles, they **rotate** the **extended thigh laterally** and **abduct** the **flexed thigh**. They also steady the femoral head in the acetabulum. The **obturator internus** covers most of the lateral wall of the **pelvis minor**. It leaves the pelvis through the lesser sciatic foramen and attaches to the greater trochanter. The narrow superior and inferior **gemelli muscles**, arising from the ischial spine and ischial tuberosity, help the obturator internus to **rotate** the **extended thigh** and to **abduct** it when it is **flexed**. The **quadratus femoris** is located inferiorly to the obturator internus and gemelli muscles. This muscle **rotates** the **thigh laterally** and steadies the femoral head in the acetabulum.

The **hip joint** is a synovial ball and socket joint between the head of the femur and the acetabulum. Movements of the joint include **flexion–extension, abduction–adduction, medial** and **lateral rotation**, and **circumduction (19)**. The depth of the acetabulum is increased by a fibrocartilaginous labrum termed the **acetabular labrum**, deepening the socket. The central and inferior parts of the acetabulum are termed the **acetabular notch**, where the **ligamentum teres passes to the fovea on the femoral head**. The joint has a **fibrous capsule**, enclosing most of the hip joint and the neck of the femur. Its synovial membrane lines the capsule of the hip joint and is reflected back along the femoral neck. The joint is innervated by branches of the femoral, sciatic and obturator nerves. The circumflex femoral arteries give the blood supply to the joint. The hip joint is stabilized by the **iliofemoral, pubofemoral** and **ischiofemoral** ligaments, and the **retinacular fibres (22)**.

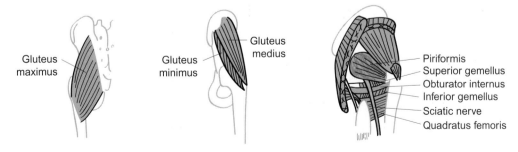

Gluteus maximus
Gluteus minimus
Gluteus medius
Piriformis
Superior gemellus
Obturator internus
Inferior gemellus
Sciatic nerve
Quadratus femoris

Answers

18. F T F F T
19. See explanation
20. 1 – A, 2 – C, 3 – B, 4 – G, 5 – D, 6 – H, 7 – E, 8 – F
21. T F T T T
22. See explanation

23. Concerning the thigh

A. Where can the femoral pulse be palpated?
B. What can be found in the adductor canal?
C. Give two purposes of the fascia lata. What makes the iliotibial tract so strong?

24. Concerning muscles in the thigh

a. The psoas major can normally be seen on an abdominal X-ray
b. The iliacus is supplied by the femoral nerve
c. Tendons of the quadriceps femoris attach to the neck of fibula
d. The sartorius is the longest muscle in the body
e. The tensor fasciae latae tightens the fascia lata

25. Concerning actions of the thigh muscles

a. The iliopsoas flexes the thigh at the hip joint
b. The sartorius is the chief flexor of the hip
c. The quadriceps femoris extends the leg at the knee joint
d. The sartorius adducts and medially rotates the thigh
e. The tensor fasciae latae helps the gluteus maximus to keep the thigh extended

26. The following diagrams show muscles of the front of the thigh, where the femoral triangle has been outlined. Label the diagrams with the options provided. Each option may be used once, more than once, or not at all

Options

A. Rectus femoris
B. Vastus lateralis
C. Adductor longus
D. Pectineus
E. Iliacus
F. Psoas minor
G. Psoas major
H. Tensor fasciae lata
I. Quadratus lumborum
J. Gracilis

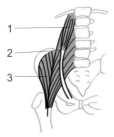

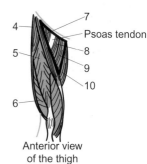

Anterior view
of the thigh

EXPLANATION: THE FEMORAL TRIANGLE AND THE ANTERIOR COMPARTMENT OF THE THIGH

The **femoral triangle** (shown in black on the figure below) is a triangular subfascial space, containing the femoral vessels and nerve. It is bounded superiorly by the **inguinal ligament**, medially by the medial border of the **adductor longus**, and laterally by the medial border of the **sartorius**. Its floor is formed (medial to lateral) by the **adductor longus**, **pectineus** and **iliopsoas**, while the roof is formed by the fasica lata (see below). The **femoral pulse** can be palpated in the triangle, at about 2–3 cm inferior to the mid-inguinal point **(23A)**.

There is an **adductor canal** which transmits the femoral vessels, lymphatic vessels and branches of the femoral nerve from the apex of the femoral triangle through the adductor hiatus to the popliteal fossa. It is bounded by the adductor longus and the adductor magnus posteriorly, and the vastus medialis medially **(23B)**.

The **superficial fascia** is a layer of **subcutaneous connective tissue** lying deep to the dermis of the skin. The superficial branches of the femoral artery, the great saphenous vein and the superficial inguinal lymph nodes can be found within this fascia. The **deep fascia (fascia lata)** is a layer of strong, dense and broad fascia covering the thigh. It provides a dense tubular sheath for the thigh muscles and prevents them from excessive bulging when contracted **(23C)**. It is attached superiorly to the inguinal ligament and the bony pelvis, and inferiorly to the tibial condyles. Its lateral part is extremely strong and is termed the **iliotibial tract**, receiving tendinous reinforcements from the **tensor fasciae latae** and **gluteus maximus**. It has a saphenous opening, a gap just inferior to the inguinal ligament in the deep fascia. The great saphenous vein courses through this gap to join the femoral vein.

Muscles in the thigh are organized into three main groups: the anterior, posterior and medial compartments (can be seen on an abdominal X-ray). The former contains the **iliopsoas**, which consists of the **psoas major** (supplied by rami of the lumbar nerves) and the **iliacus** (supplied by the femoral nerve). The psoas major runs from the lumbar vertebrae to the lesser trochanter of the femur, while the **iliacus** lies along the lateral side of the psoas major. They are the prime flexors and medial rotators of the thigh at the hip joint. The **tensor fasciae latae** lies laterally and is enclosed between the two layers of the fascia lata. This tightens the fascia lata. Innervated by the superior gluteal nerve, it helps the **gluteus maximus** to maintain the extended knee.

The **sartorius** is the longest muscle in the body and is the most superficial muscle of the anterior thigh. It covers the femoral artery and runs in the adductor canal. Supplied by the femoral nerve, it flexes, abducts and laterally rotates the thigh at the hip joint. The **quadriceps femoris** is supplied by the femoral nerve and it extends the leg at the knee joint. The muscle is divided into four parts: the **rectus femoris**, **vastus lateralis**, **vastus medialis** and **vastus intermedius**. Their tendons unite to form the **quadriceps tendon**, which attaches to and surround the patella and also the tibial tuberosity.

In addition, the anterior compartment contains the **femoral artery, femoral vein** and **femoral nerve**.

Answers

23. See explanation
24. T T F T T
25. T F T F F
26. 1 – F, 2 – G, 3 – E, 4 – H, 5 – A, 6 – B, 7 – E, 8 – D, 9 – C, 10 – J

27. Concerning the muscles of the medial compartment of the thigh

    a. They adduct the thigh at the hip joint
    b. They are all supplied by the femoral nerve
    c. The adductor longus is the largest in the group
    d. The adductor brevis is the weakest adductor of the group
    e. The tendon of the obturator externus crosses the posterior part of the neck of the femur

28. Concerning muscles in the medial compartment of the thigh, label the following diagram with the options provided

Options

    A. Adductor longus
    B. Gracilis
    C. Adductor brevis
    D. Obturator externus
    E. Adductor magnus
    F. Pectineus

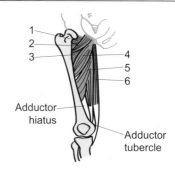

29. Concerning the hamstrings

    a. They all attach proximally to the ischial tuberosity
    b. They are all supplied by the femoral nerve
    c. The semitendinosus muscle laterally rotates the tibia on the femur with the semimembranosus muscle
    d. Their chief action is to extend the leg
    e. The long head of the biceps femoris rotates the leg laterally

30. The following diagram shows the posterior compartment of the thigh. Label the diagram with the options provided

Options

    A. Semitendinosus
    B. Semimembranosus
    C. Biceps femoris

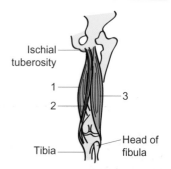

EXPLANATION: THE MEDIAL AND POSTERIOR COMPARTMENTS OF THE THIGH

The **medial compartment** of the thigh is mainly involved with the **adduction** of the thigh. All muscles in this compartment are supplied by the **obturator nerve**, except for the pectineus (femoral nerve). The compartment contains the following muscles:

- **Pectineus:**
- **Adductor longus:** is a long adductor lying in the most anterior aspect of the adductor group
- **Adductor brevis:** lies deep to the pectineus and adductor longus
- **Adductor magnus:** is the largest muscle in the adductor group – consists of **adductor** and **hamstring** parts
- **Gracilis:** lies on the medial side of the thigh and knee. It is the most superficial adductor of the compartment and also the weakest. This is why surgeons often transplant this muscle to replace a damaged one
- **Obturator externus:** is the deepest adductor, with its tendon crossing the posterior aspect of the neck of the femur.

Furthermore, the **profunda femoris artery**, its medial circumflex femoral and perforating branches and the obturator artery all lie within this compartment. The **profunda femoris** and **obturator veins** can also be found here, as well as the **obturator nerve**.

The **posterior compartment** of the thigh contains the **hamstrings**. They all attach proximally to the **ischial tuberosity** and are all supplied by the tibial nerve. They **extend** the **thigh** and **flex** the **leg**. The hamstrings consist of three large muscles:

- **Semitendinosus:** is a half-tendinous muscle
- **Semimembranosus:** a half-membranous muscle. Together with the semitendinosus, it has an additional action of rotating the tibia on the femur medially
- **Biceps femoris:** is a two-headed (long and short) muscle and has an additional attachment to the body of the femur. It **flexes** the **leg** and **rotates** it **laterally**. Its long head also **extends** the **thigh**.

In addition, the posterior compartment contains the perforating branches of the **profunda femoris**, **venae comitantes** of the small arteries and the **sciatic nerve**.

Answers

27. T F F F T
28. 1 – D, 2 – F, 3 – C, 4 – A, 5 – E, 6 – B
29. T F F F T
30. 1 – B, 2 – A, 3 – C

31. Concerning the knee joint

a. It is a synovial hinge joint
b. The anterior cruciate ligament prevents slipping of the femur forward on the tibia
c. The posterior cruciate ligament inserts onto the lateral surface of the medial condyle of the femur
d. The collateral ligaments are taut on full extension of the knee joint
e. The lateral collateral ligament has deep fibres firmly attached to the lateral meniscus

32. Concerning the knee

a. The oblique popliteal ligament strengthens the fibrous capsule of the knee posteriorly
b. The knee joint has five bursae that communicate with the synovial cavity of the knee joint
c. The lateral meniscus is firmly attached to the deep surface of the tibial collateral ligament
d. The menisci are commonly torn by the rotation of an extended knee
e. A pulse can be felt in the popliteal fossa

33. The diagram on the left shows an anterior view of the knee joint and the one on the right shows its posterior view. Label them with the options provided. You may use the options once, more than once, or not at all

Options

A. Lateral collateral ligament
B. Medial collateral ligament
C. Transverse ligament of knee
D. Posterior cruciate ligament
E. Anterior cruciate ligament

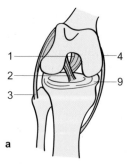

a

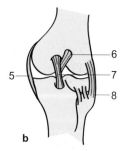

b

34. What forms the boundaries of the popliteal fossa?

EXPLANATION: THE KNEE AND THE POPLITEAL FOSSA

The **knee joint** is a **synovial hinge joint**, permitting a small degree of rotation. The joint mainly allows **flexion** and **extension** of the knee. It is supplied by the genicular branches of the **popliteal artery** and is innervated by branches of the **femoral**, **tibial** and **common peroneal** and **obturator nerves**. It consists of three articulations: between the patella and femur (patello-femoral) and the medial lateral and condyles of the femur and tibia. The knee joint has a strong **fibrous capsule**, attaching to the margins of the articular surfaces. The joint is protected by ligaments. These can be divided into the **intracapsular** and **extracapsular ligaments**.

The **intracapsular ligaments** include the **cruciate ligaments,** which are intracapsular but extra-synovial. There are two cruciate ligaments:
- **Anterior cruciate ligament:** arises from the anterior part of the intercondylar region and inserts onto the medial surface of the lateral condyle of the femur. It prevents hyperextension of the knee and the forward movement of the tibia on the femur
- **Posterior cruciate ligament:** arises from the posterior intercondylar ridge of the tibia and inserts onto the lateral surface of the medial condyle of the femur. It **prevents slipping** of the **femur forward** on the tibia.

The **extracapsular ligaments** include:
- **Collateral ligaments**: the **lateral collateral ligament** extends from the lateral epicondyle of the femur to the lateral surface of the head of the fibula, while the **medial collateral ligament** extends from the medial epicondyle of the femur to the medial surface of the tibia. The latter has deep fibres attached to the medial meniscus and the fibrous capsule of the knee joint. Both ligaments are taut on full extension of the joint
- **Oblique popliteal ligament:** is found in the back of the knee and strengthens the capsule posteriorly
- **Arcuate popliteal ligament:** arises from the posterior aspect of the head of the fibula and spreads out over the back of the knee joint, inserting into the intercondylar region of the tibia and the posterior region of the lateral epicondyle of the femur.

The incongruence of the articular surfaces is compensated for by the presence of articular fibrocartilaginous pads (**menisci**). These 'shock absorbers' of the joint are attached to the tibial intercondylar region and to each other via the transverse (genicular) ligament. The **medial meniscus** is C-shaped and is firmly attached to the deep surface of the tibial collateral ligament. The **lateral meniscus** is loosely attached to the tibia and to the femur by two meniscofemoral ligaments. They are commonly torn by extreme flexion–rotation of the knee, or rotation of the extended knee. The knee joint has five bursae that communicate with the synovial cavity of the knee joint.

The popliteal fossa is a diamond-shaped region posterior to the knee. Its boundaries are as follows: the roof of the popliteal fossa consists of the deep fascia, penetrated by the small saphenous vein; the biceps femoris forms the superolateral border, and the semitendinosus muscle forms the superomedial border; both the inferomedial and inferolateral borders are formed by the heads of the gastrocnemius (34). The fossa contains, from deep to superficial, the popliteal artery, popliteal vein and tibial nerve. The common peroneal nerve can also be found in the fossa, running along the medial border of the biceps tendon.

Answers
31. T F T T F
32. T T F T T
33. 1 – D, 2 – E, 3 – A, 4 – B, 5 – B, 6 – E, 7 – D, 8 – A, 9 – C
34. See explanation

35. Concerning the leg

a. The crural fascia is fused with superficial peroneal periosteum of the tibia
b. The crural fascia is continuous with the fascia lata
c. The peroneus longus is innervated by the nerve
d. Muscles in the posterior compartment are supplied by the peroneal artery
e. The extensor hallucis longus is innervated by the deep peroneal nerve

36. Concerning muscles in the leg

a. The extensor digitorum plantarflexes the foot
b. The gastrocnemius is innervated by the tibial nerve
c. The peroneus longus everts the foot
d. The superficial flexor muscles attach to the calcaneus via the tendocalcaneus
e. The tibialis posterior is the prime flexor of the big toe

37. The following diagrams show muscles in the leg. Label the diagrams with the options provided. Each option may be used once or more than once

Options

A. Gastrocnemius
D. Extensor digitorum longus
G. Peroneus longus

B. Extensor retinaculum
E. Tibialis anterior

C. Soleus
F. Peroneus brevis

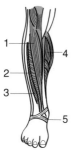

Anterior compartment

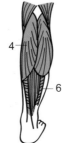

Posterior compartment

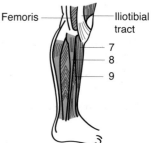

Lateral compartment

EXPLANATION: THE LEG

The leg has a deep fascia (**crural fascia**) that is continuous with the fascia lata of the thigh. It is attached to the anterior and medial borders of the tibia and fused with its periosteum. From the deep surface of the crural fascia, the **intermuscular septa** divide the leg into three compartments. At the ankle, the deep fascia forms retinaculae for the tendons of these compartments.

The **anterior (extensor) compartment** of the leg is supplied by **the anterior tibial artery** and is innervated by the **deep fibular nerve** of the sciatic nerve. The compartment consists of four muscles:
- **Tibialis anterior:** dorsiflexes and inverts the foot
- **Extensor hallucis:** extends the hallux and dorsiflexes the foot
- **Extensor digitorum:** extends the lateral four digits and dorsiflexes the foot
- **Fibularis tertius:** dorsiflexes the foot. Also helps out in eversion of the foot

The **lateral compartment** of the leg is innervated by the **superficial fibular nerve**. There are no arteries in this compartment except for branches to the fibular muscles from the fibular artery. This compartment consists of two muscles: **fibularis longus** and **fibularis brevis**. Both muscles **evert the foot** and **weakly plantar flex it**.

The **posterior (flexor) compartment** of the leg is supplied by **the posterior tibial artery** and is innervated by the **tibial nerve**. Muscles in this compartment can be considered in two groups:

- The **superficial group** insert into the middle third of the posterior surface of the calcaneus via the tendocalcaneus (**Achilles tendon**). They act as a muscle pump, squeezing venous blood upwards during contraction. These muscles include:
  - **Gastrocnemius:** forming the prominence of the calf, it plantarflexes the foot and flexes the knee joint. Contraction of this muscle produces rapid movement during running and jumping
  - **Soleus:** lying deep to the gastrocnemius, the soleus plantarflexes the foot, steadies the leg on foot and prevents falling anteriorly when standing

- The **deep group** consists of four muscles:
  - **Popliteus:** unlocks the extended leg by rotating the femur laterally on the fixed tibia
  - **Flexor hallucis longus:** flexes the hallux and plantarflexes the foot
  - **Tibialis posterior:** plantarflexes and inverts foot
  - **Flexor digitorum longus:** flexor of the lateral four digits

Answers

35. T T T F T
36. F F T T F
37. 1 – E, 2 – D, 3 – G, 4 – A, 5 – B, 6 – C, 7 – E, 8 – G, 9 – F

38. The ankle joint

    a. Is a synovial hinge joint
    b. Permits dorsiflexion and plantarflexion
    c. Has two bursae
    d. Is innervated by the common peroneal nerve
    e. Is particularly stable during dorsiflexion

39. Concerning the foot

    a. The midtarsal joint permits inversion–eversion movements of the foot
    b. Intermetatarsal joints are synovial condyloid joints
    c. The abductor hallucis is supplied by the medial plantar nerve
    d. The plantar interossei muscles abduct the digits of the foot
    e. The flexor digitorum brevis flexes the lateral four digits of the foot

40. The following diagrams are muscles of the foot. Label them with the options provided. Which nerve supplies structure 1?

Options

    A. Quadratus plantae   **B.** Flexor digiti minimi brevis   **C.** Flexor digitorum brevis
    D. Abductor hallucis   **E.** Adductor hallucis   **F.** Lumbricals
    G. Abductor digiti minimi

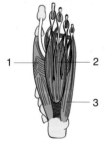

First layer of plantar muscles

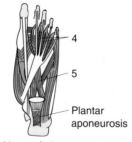

Second layer of plantar muscles

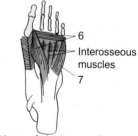

Third layer of plantar muscles

EXPLANATION: THE ANKLE AND THE FOOT

The **ankle joint** is a **synovial hinge joint**, where the inferior ends of the tibia and fibula form a mortise articulating with the trochlea of the talus. The joint permits **dorsiflexion** and **plantarflexion**. However, if the foot is plantarflexed, a certain degree of rotation, abduction and adduction can be performed. The ankle joint is particularly **stable during dorsiflexion** due to increased articulation of the bones. The joint has a fibrous capsule enclosing the articular surfaces, supported by strong ligaments. These include the medial and lateral ligaments of the ankle. The joint has a synovium but has **no bursae**. It is innervated by the **deep fibular and tibial nerves**, and receives blood supply from the fibular, and anterior and posterior tibial arteries.

The **foot joints** consist of:

- **Subtalar joint:** the talus rests on and articulates with the calcaneus at these synovial joints. It permits inversion and eversion of the foot.
- **Midtarsal joint:** composed of the calcaneocuboid and talomavicular joints, the midtarsal joint is involved with foot inversion–eversion movements
- Other foot joints include the **tarsal, tarsometatarsal** (synovial plane), **intermetatarsal** (synovial plane), **metatarsophalangeal** (synovial condyloid) and **interphalangeal** (synovial hinge) joints.

In the **muscles** of the **foot**, the **medial plantar nerve** supplies the abductor hallucis, flexor hallucis brevis, first lumbrical, and cutaneously the medial three-and-a-half toes and the sole. All other intrinsic muscles and skin overlying the lateral one-and-a-half toes and sole are supplied by the **lateral plantar nerve**. These muscles are divided into four layers:

- **First layer** consisting of the **abductor hallucis** (abducts and flexes the big toe [hallux]), **flexor digitorum brevis** (flexes the lateral four digits) and **abductor digiti minimi** (abducts and flexes the fifth digit)
- **Second layer** consisting of the **quadratus plantae**, which assists the flexor digitorum longus to flex the lateral four digits, and the **lumbricals**, which flex the proximal phalanges and extend the middle and distal phalanges of the lateral four digits
- **Third layer** consisting of the **flexor hallucis brevis** (flexes the proximal phalanx of the big toe), **adductor hallucis** (adducts the big toe) and the **flexor digiti minimi brevis** (flexes the proximal phalanx of the fifth digit). They lie in the anterior half of the sole of the foot
- **Fourth layer** consisting of the **interosseous muscles**, occupying the spaces between the metatarsal bones. The **p**lantar interossei **ad**duct the digits (**PAD**) and the **d**orsal interossei **ab**duct (**DAB**) the digits and flex the metatarsophalangeal joints.

Answers
38. T T F F T
39. T F T F T
40. 1 – D, 2 – C, 3 – G, 4 – F, 5 – A, 6 – E, 7 – B, medial plantar nerve

41. Case study

A 64-year-old woman slips on the floor in her bathroom. She is admitted to the emergency room with severe pain and is unable to walk. On examination her left lower limb is rotated laterally and looks shorter than her right lower limb.

   A. What is the likely diagnosis?
   B. What complications may be seen?

42. Case study

A 65-year-old man presents to his GP with pain in his right buttock, which spreads to the posterior region of his right leg. While the patient is lying down, he is unable to raise his right leg when extended fully due to pain.

   A. What is the likely diagnosis?

43. Case study

A 60-year-old man presents to his GP with a gripping, tight, cramp-like pain in his right calf muscles on exercise. The pain disappears on resting. On examination the femoral pulse is present, but the popliteal, posterior tibial and dorsalis pedis pulses are absent in the right lower limb.

   A. What is the likely diagnosis?
   B. Where is the lesion likely to be?

44. Case study

A 29-year-old woman in the third trimester of her pregnancy presents to her GP with noticeable dilated and tortuous veins in her lower extremities. The veins are even more noticeable after prolonged standing.

   A. What is the likely diagnosis?
   B. What are the complications of the disease? Does the patient require prophylactic measurements to prevent deep vein thrombosis and pulmonary embolism?

GP, general practitioner

EXPLANATION: CLINICAL SCENARIOS

41.

This woman is likely to have a **fractured femur**. As discussed on page 17, postmenopausal women are at risk of developing osteoporosis, leading to a decreased total bone mass. The most fragile part of the femur is the neck. She may have exerted a torsional force on one hip whilst slipping, fracturing the femoral neck in the process (i.e. the fracture was the cause of the fall, not as a result of the fall). Due to the fact that branches of the circumflex femoral arteries may be damaged with the fracture, blood supply may be diminished, leading to **avascular necrosis** of the femoral head (bone death due to poor blood supply).

42.

This man may have damaged his **sciatic nerve**, giving him signs of **sciatica**. The nerve extends from the pelvis to the leg and foot, so damage leads to pain in that region. Straight leg raising stretches the sciatic nerve and leads to more pain. A herniated intervertebral disc is a common cause of sciatica.

43.

This patient presents with symptoms of **intermittent claudication**, which is pain occurring with walking. The pain typically appears after walking for a certain distance and disappears after resting. It is a marker for generalized **atherosclerotic occlusive disease**. As the femoral pulse can be felt, whereas pulses cannot be felt from the popliteal downwards, the lesion is likely to be between the groin and the knee.

44.

This woman presents with **varicose veins**. They are dilated, tortuous veins caused by increased intraluminal pressure affecting the superficial veins of the lower extremities. It is common in females and in pregnancy. It is aggravated by prolonged standing or sitting. Complications include: oedema, thrombosis, stasis dermatitis and ulcerations. However, it is rarely a source of emboli.

Answers

41. See explanation
42. See explanation
43. See explanation
44. See explanation

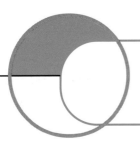

INDEX